Telecourse Guide for

THE SOCIOLOGICAL IMAGINATION:

Introduction to Sociology

Fourth Edition

Telecourse Guide for

THE SOCIOLOGICAL IMAGINATION:

Introduction to Sociology

Fourth Edition

Glenn Currier
Jane Penney

Produced by:

DALLAS TELELEARNING
Dallas County Community College District

HARCOURT BRACE COLLEGE PUBLISHERS
Fort Worth Philadelphia San Diego New York Orlando Austin San Antonio
Toronto Montreal London Sydney Tokyo

This telecourse guide is the work of many contributors, but Betsy Turner, former Assistant Dean, Distance Education Services, and Evelyn Wong, Telecommunications Information Specialist, were always there for me, somehow translating my scribbled, "sticky" note pad messages with clarity and understanding while providing guidance and suggestions. Tracey McKenzie, sociologist, and Karen Pate, instructional designer, were invaluable in helping produce this telecourse guide and making the revisions necessary for the telecourse guide to be ready for student use in the fall of 1999. I am indebted to Glenn Currier for laying a fine foundation on which to grow; he truly was the "pioneer" in the distance learning sociology course and has proven to be a valuable and enduring friend throughout the years. The Dallas County Community College District's R. Jan LeCroy Center for Educational Telecommunications is filled with individuals who possess the quickly changing vision of learning as a guiding light... Pamela K. Quinn, Assistant Chancellor/LeCroy Center; Jacquelyn B. Tulloch, Executive Dean, Distance Education; and Nora Coto Busby, former Instructional Designer, plus many individuals who truly keep the whole thing working, including Liz Gonzalez and Millie Davidson.

Above all, however, I want to thank my students who through the years have contributed to my personal growth and my dedication to education. My philosophy of education has changed through the years, and I have my students to thank for my becoming a better, continually changing educator. I am sure I have learned more from them than I imparted to them! I hope students have left my class with a renewed sense of empowerment enabling them to take charge of learning throughout their lives...inside and outside of any "classroom"!!

ISBN: 0-15-512856-6

Address for editorial correspondence:
Harcourt Brace College Publishers
301 Commerce Street, Suite 3700
Fort Worth, TX 76102

This edition has been printed directly from camera-ready copy. Photographs reproduced herein were supplied by the Dallas County Community College District archives.

Printed in the United States of America
6 7 8 9 0 1 2 3 4 5 000 10 9 8 7 6 5 4 3 2 1

Contents

To You, the Student

Education is not the filling of a pail, but the lighting of a fire.
—William Butler Yeats

The ultimate goal of the educational system is to shift to the individual the burden of pursuing his education.
—John W. Gardner

The quotations of William Butler Yeats and John W. Gardner encompass the heart of my teaching philosophy which has evolved over many years in the classroom. When I first began teaching, I had the illusion that "I" magically imparted the knowledge of sociology to my students who then had the responsibility of somehow grasping the information. And, of course, everything in the discipline of sociology was "necessary" for my students to comprehend, and tests over this "necessary" information reinforced the vital importance of memorizing all those terms, concepts, theories, and names.

How my role as a sociology professor has changed!! It didn't take long for me to discover a key of education: making those terms, concepts, theories, and names relevant to my students' lives; without that relevancy, students learn (memorize) but quickly forget and do not gain the ability to apply sociology in their own lives. Educators must ignite the fire and desire to learn within students. We must shift the responsibility of learning to students through student empowerment which drives the desire to learn because the information helps students deal with everyday events. To gain the ability to take what one learns in the classroom and apply such knowledge and awareness to one's life is an important ingredient in learning.

Education does not occur in a vacuum nor only in the traditional classroom with the students being "fed" information by the instructor. Today my "classroom" has been redefined through technological changes enabling courses to be available regardless of distance and barriers. Students are connected through such technology as computers, voice bridges, interactive video, plus a multitude of combinations. Today, thousands of students watch televised programs in the comfort of their own homes. Students have the capability, with technological support, to connect to classmates all over the world. Yes, the "classroom" has taken on a new definition and scope.

It is my hope that you will ask yourself how the information in your sociology course applies to your life. Education must build a bridge between the "classroom" (however defined) and your life. Students and educators alike share in this task! Begin today to ask yourself how sociology applies in your life, empowering you with the tools to better understand the world and human differences. Our world has grown smaller, and you will need a "sociological imagination" to fully understand and appreciate your personal life as well as our global community. Enjoy your journey into the wonders of sociology!!!

—Jane Penney

Course Organization

The Sociological Imagination is designed as a comprehensive learning system consisting of three elements: telecourse guide, textbook, and video programs.

TELECOURSE GUIDE

The telecourse guide for this course is:

Currier, Glenn, and Jane Penney. *Telecourse Guide for The Sociological Imagination: Introduction to Sociology,* 4th ed. Fort Worth, Texas: Harcourt Brace College Publishers, 1999.

This telecourse guide acts as your daily instructor. For each lesson it gives you a lesson assignment; an overview; lesson goals; textbook and video objectives; reader objectives when applicable; related activities; and a practice test. If you follow the telecourse guide recommendations and view each lesson carefully, you should successfully accomplish all of the requirements for this course.

TEXTBOOK

In addition to the telecourse guide, the textbook required for this course is:

Kornblum, William, *Sociology in a Changing World,* 5th ed. Fort Worth, Texas: Harcourt Brace College Publishers, 1999.

William Kornblum, an active researcher, writer, and teacher presents sociology as a dynamic social science that is involved in the social world. He does not focus on definitions and terminology for their own sake, but approaches sociology more as a process of inquiry into the social world and as a means of understanding and dealing with social change. This approach not only gives you a different perspective on historical patterns but also helps you to understand social processes, conflicts, and changes occurring today.

VIDEO PROGRAMS

The video program series for this course is:

The Sociological Imagination

Each video program is correlated to a specific reading assignment for that lesson. The video programs are packed with information, so watch them closely. If the lessons are broadcast more than once in your area, or if video or audio tapes are available at your college, you might find it helpful to watch the video programs again or listen to audio tapes for review. Since examination questions will be taken from the video programs as well as from the textbook, careful attention to both is vital to your success.

Telecourse Guidelines

Follow these guidelines as you study the material presented in each lesson:

1. LESSON ASSIGNMENT:
   Review the Lesson Assignment in order to schedule your time appropriately. Pay careful attention; the titles and numbers of the textbook chapter, the telecourse guide lesson, and the video program may be different from one another.

2. OVERVIEW:
   Read the Overview for an introduction to the lesson material.

3. LESSON GOALS:
   Review the Lesson Goals and pay particular attention to the lesson material that relates to them.

4. TEXTBOOK OBJECTIVES:
   To get the most from your reading, review the textbook objectives, then read the assignment. You may want to write responses or notes to reinforce what you have learned.

5. READING OBJECTIVES:
   Readings are included in some lessons of the telecourse guide. To get the most from your reading, review the reading objectives, then read the reading. You may want to write responses or notes to reinforce what you have learned.

6. VIDEO OBJECTIVES:
   To get the most from the video segment of the lesson, review the video objectives, then watch the video. You may want to write responses or notes to reinforce what you have learned.

7. RELATED ACTIVITIES:
   These activities may be used by your instructor as written assignments or as discussion topics. They may also be included as essay questions on your tests.

8. PRACTICE TEST:
   Complete the Practice Test to help you evaluate your understanding of the lesson.

9. ANSWER KEY:
   Use the Answer Key at the end of the lesson to check your answers or to locate material related to each question of the Practice Test.

Notes and Assignments:

Lesson 1

From Social Interaction to Social Structure

LESSON ASSIGNMENT

Review the following assignment in order to schedule your time appropriately. Pay careful attention; the titles and numbers of the textbook chapter, the telecourse guide, and the video program may be different from one another.

Text:

Kornblum, *Sociology in a Changing World,*
Chapter 1, "Sociology: An Introduction," pp. 2-17.

Reading #1:

Telecourse Guide for The Sociological Imagination, pp. 5-7.

Reading #2:

Telecourse Guide for The Sociological Imagination, pp. 8-9.

Video:

"From Social Interaction to Social Structure,"
From the series, *The Sociological Imagination.*

OVERVIEW

Roberta and Greg met C. Wright Mills at the same place they met each other—in an introductory sociology class. Although they encountered Mills only in their text book and in the words of their teacher, Mills' ideas piqued their interest. Some of those ideas, Roberta and Greg agreed, were eye-opening and insightful; on others, their opinions differed.

In fact, it was Mills who provoked their first encounter, as they engaged in a brief exchange across the classroom—to the delight of the teacher and the other students. Even as Greg and Roberta argued, however, they found themselves instantly attracted to each other. Her sparkling eyes and knowing smile, his sincerity and compassion—each produced a strong response in the other. They couldn't resist continuing their discussion in the little Italian restaurant near the campus.

That became their favorite meeting place after every sociology class, as Greg and Roberta became almost as enchanted with sociology as with each other. In fact, one sociological subject or another continued to stimulate them to spirited dialogue, even as—in later years—their children tugged at their clothing. Little Wright pestered his mother, and Arlie charmed her father. Both children were named after sociologists the parents had come to admire in their college courses.

We hope this encounter with sociology will be as intellectually fruitful for you as it was for Roberta and Greg. Even if you do not build a long-term relationship with a fellow student as a result of this class, we expect that you will profit immensely from your decision to put your time and energy into learning more about yourself and society.

In fact, that is what sociology is about: the study of social behavior. You and your friends and family comprise numerous social groups. On another level of social behavior, you pay taxes and interact in a multiplicity of ways within your community or society; therefore, you are part of those social systems as well.

Perhaps without realizing it, you already know a great deal about the subjects you will be studying in this course. The more you have interacted in various groups and social settings during your lifetime, the more familiar you already are with the subjects you will be learning about here.

Much of your knowledge of social behavior until now, however, is based on what you have experienced as an individual person. What we hope to do in this course is to broaden that experiential knowledge and transform it into intellectual knowledge. That will be even more useful to you as you confront new and different situations throughout your lifetime.

As you learn how sociologists have used the science of sociology to make sense of the social world, you will develop intellectual tools that will help you to function better in that world. The ideas and perspectives you acquire here will let you stand back a little from your individual circumstances and see the web of your own social relationships in a new light. As a result, you will find yourself freer, because you will be less subject to control by your current social situation and less likely to blame yourself for those aspects of your life that arise from social forces beyond you.

This new vision and awareness is called a "sociological imagination." In fact, the reason we named this course *The Sociological Imagination* is because that is precisely what we expect you to gain from it.

C. Wright Mills is the sociologist who invented this term. His perspicacity, combined with his discomfort with the world and with his own field of study, drove him to propose this uncommon perspective on life, which he dubbed the sociological imagination. Just as Greg and Roberta were hooked by its promise, we hope that you too will come to new realizations about the intersection of your personal biography with the social worlds that surround you.

We can't delay introducing you to these ideas. From the start, we feel impelled to awaken you to the fascination of sociological thinking. Therefore, the Readings in the first two lessons introduce you to the basic concepts of sociology.

The smallest possible piece of social behavior that sociologists study is a single human interaction. Our social relationships, groups, communities, and societies would not be possible without social interaction. Have you ever tried to have a relationship with someone who refused to interact with you? It is hopeless. In a sense, sociology is the study of social interaction, so we begin this course with an examination of human social interaction as the basis of social life.

Beyond the micro-level of our lives are many larger social forms and forces that affect us—our society, for example. Sociologists are fascinated with human social life, at whatever level it takes place—from social interaction to the formation of groups, communities, and societies. Perhaps this is because we sociologists are curious people, even nosy. We can't resist snooping around churches, families, and even bars and brothels. Wherever human beings are interacting, sociologists explore, observe, and record their findings.

If you too are curious and can't stop yourself from eavesdropping on the conversation at the table next to you, if you yearn to know about the human stories beyond the headlines and your taken-for-granted world, we invite you to join us on this captivating intellectual journey into *The Sociological Imagination*.

LESSON GOALS

Upon completing this lesson, you should be able to:

1. Analyze the sociological imagination and sociology—how sociology developed and the different levels at which it analyzes social reality.

2. Analyze how the following concepts are related to each other: social interaction, culture, and socialization.

3. Analyze the different levels and kinds of social structures including social groups, communities, and societies.

4. Analyze how social interaction leads to social relationships, as well as how these relationships combine to form social structures.

TEXTBOOK OBJECTIVES

The following textbook objectives are designed to help you get the most from the text. Review them, then read the assignment. You may want to write notes to reinforce what you have learned.

Text: Kornblum, *Sociology in a Changing World*, Chapter 1.

1. Explain the meaning of the sociological imagination. Describe and exemplify one of the main objectives of the textbook.

2. Define sociology. Describe the aspects of the social environment sociologists study.

3. Describe and exemplify the levels of social reality analyzed by sociologists.

READING #1
SOCIAL INTERACTION, CULTURE, AND SOCIALIZATION

READING OBJECTIVES

The following reading objectives are designed to help you get the most from Reading #1. Review them, then read the assignment. You may want to write notes to reinforce what you have learned.

Reading #1: *Telecourse Guide for The Sociological Imagination*, pp. 5-7.

1. Recognize examples of social (or symbolic) interaction.
2. Explain and exemplify the importance of culture and language in social interaction.
3. Explain the meaning of socialization.
4. Describe the role of social interaction in the socialization process.
5. Define culture and explain what culture includes.
6. Identify ways that culture is passed down from one generation to the next.

The Meaning and Significance of Social Interaction

Social interaction is the process by which we mutually influence each other in our actions. In the video program, you saw and heard two people going through this reciprocal process. Greg and Roberta also engaged in the process of interaction, both in their sociology class and later at the restaurant. Remember, it is not enough for one person to deliver a message. Unless the other person both reacts and responds to that message, no interaction takes place.

You also learned that social interaction is symbolic: it involves the exchange of messages that have meaning. Part of the meaning communicated in interaction is influence by the culture of those communicating.

Roberta defined Greg not merely as a human being that happened to be in her presence, but as a "man," since Greg possessed certain traits delineated by her culture. Her reaction and her messages to him would have been quite different had he been a woman.

Every culture has its own concept of what it means to be a man or a woman. Those concepts, their meanings, and the language that expresses those concepts are carried in the minds of the members of a society and communicated in the course of their interaction.

For example, feminists are offended by the cultural definition of woman as "sex object." They also dislike language that is sexist. Both of these—concepts (or meanings) and language (symbols expressing the concepts)—are part of the culture that feminists want to change. And these meanings are often communicated through language and other symbols as part of the social interactions between the two sexes.

Of course, other culturally patterned concepts are communicated as well. Roberta and Greg understood themselves to be both "students" and "adults," and these definitions also influenced the content and direction of their messages and actions.

Culture is very important in every social interaction. *Culture* is all the modes of thought, behavior, and production that are handed down from one generation to the next by means of communicative interaction—through speech, gestures, writing, building, and all other communication among humans—rather than by genetic transmission or heredity. Culture is the way of life of a society. It includes the society's ideas, customs, rules (norms), knowledge, values, and material objects. All these aspects of a society's culture are reflected in its language. Not only does culture affect interactions, but the opposite also happens. People actually create and change the culture as they interact. For example, every year new editions of dictionaries are published with additional words that have come into use as people both cause and react to changing situations. Just as interaction is a process, so is culture.

You know how you can be affected by what others say to you. You may feel hurt, afraid, or angry, or their words may elicit feelings of pride, joy, or satisfaction. Sociologists have studied how adults influence each other in the course of their symbolic interaction. But sociologists also have paid particular attention to the impact of interaction on children.

Studies have found the interaction process to be vital to children in the formation of their self-image or "self." As children interact, they take on the roles and expectations communicated to them by adults or other children. For instance, they learn what it means to be a child, to be a boy or girl, and they play those roles.

Meanwhile, the messages children are constantly receiving influence the level of self-esteem they develop. If the messages label them as "bad" or "less than...," the children may grow up feeling inferior. This process of forming personality and self-image, and of learning how to be a social person, is called *socialization*. Becoming a social person means that a person learns to conform to society's norms, values, and roles.

READING #2
SOCIAL RELATIONSHIPS AND SOCIAL STRUCTURE

READING OBJECTIVES

The following reading objectives are designed to help you get the most from Reading #2. Review them, then read the assignment. You may want to write notes to reinforce what you have learned.

Reading #2: *Telecourse Guide for The Sociological Imagination*, pp. 8-9.

1. Explain the meaning of a social relationship. Identify different social relationships.
2. Define a social structure, and identify the levels or kinds of social structures.
3. Recognize examples of groups.
4. Describe what a community is and the different kinds of communities that exist. Be able to recognize examples.
5. Define society and recognize examples of society.

Social Structure—A Web of Relationships

You can see that a single social interaction is very complex, yet one interaction does not make a relationship. You may interact with someone on the street and never see that person again. In that case, interaction occurred, but no relationship developed.

A *social relationship* is based on a pattern of interactions. You probably know from experience that your personal relationships take time and effort to establish and maintain. Long-term committed relationships, such as marriage, require determined effort to interact and communicate. Indeed, a major cause of divorce is problems in communication. Only after interacting on a regular basis, over a period of time, can you build the pattern of interaction necessary for a personal relationship.

Some relationships are impersonal, yet they are still based on patterned interaction. For example, teachers and students have been interacting for centuries. The teacher-student relationship of today is based, at least partially, on the pattern of those teacher-student interactions built up over a long period of time.

In the video program, you see examples of *social structure*: the web of relationships that people create as they interact. According to sociologist William Kornblum, *social structure* refers to recurring patterns of behavior that people create through their interactions and relationships. There are many levels of social structure, from a group to a society.

A *group* is any collection of people who interact on the basis of shared expectations regarding one another's behavior. A family, a softball team, a set of workers at a restaurant who work closely together are called primary groups. Primary groups are usually small, interaction is face-to-face, and the relationships are personal and close. A large corporation is an example of a secondary group. Interaction is more impersonal, people relate to each other in specified roles such as president, division head, and line worker.

Another type of social structure is a community. William Kornblum states that *communities* are sets of primary and secondary groups in which the individual carries out important life functions such as raising a family, gaining a living, finding shelter, and the like. Communities may be either territorial or non-territorial. Both include primary and secondary groups, but territorial communities are contained within geographic boundaries, whereas non-territorial communities are networks of associations that form around shared goals. A neighborhood is a territorial community; a "professional community" is a non-territorial community.

The largest social structure to which any of us belong is a *society*. A society is a population of people that is organized in a cooperative manner to carry out the major functions of life, including reproduction, sustenance, shelter, defense, and disposal of the dead. The term population refers to a number of people in a certain region. A population is not a social structure. To be a society, a population must be organized to the extent that there is a web of social relationships with recurring patterns of behavior. The idea of a society stresses the interrelationships among people.

In summary, people interact and form relationships. When those relationships become coordinated or organized, they are a social structure. There are many types or levels of social structures: groups, including primary and secondary groups, communities, and the largest social structure, a society.

Social structure is a central concept of sociology and throughout this course, you will learn more about how you are affected by numerous social structures in which you live, work, and play.

VIDEO OBJECTIVES

The following video objectives are designed to help you get the most from the video segment of this lesson. Review them, then watch the video. You may want to write notes to reinforce what you have learned.

Video: "From Social Interaction to Social Structure"

1. Define sociology. Explain the meaning of social interaction.

2. Explain how interaction becomes a social relationship and how relationships become social structures. Describe examples of this process given in the video program. Explain how the video program defines and exemplifies a social structure.

3. Explain the importance of roles both within and outside of the family.

4. Identify the different levels sociologists study social structures.

RELATED ACTIVITIES

These activities may be used by your instructor as written assignments or as discussion topics. They may also be included as essay questions on your tests.

1. Can you remember an incident in which someone important to you—a parent, relative, friend, or teacher—spoke to you in a manner that made you feel elevated and respected? Did this interaction influence your sense of "self" in a positive way? Explain. Report the incident and your conclusions in a brief narrative.

2. List five things, material or non-material items, that are part of your culture. Keep your list brief; explain the meaning of each item; a line or two will do for each item. For example: People on a submarine, on a ship, in a remote location, or in the military can constitute a culture of their own. What material or non-material items are part of that culture?

3. List five social structures to which you belong. In your list, include at least two different levels of social structures.

4. In a brief narrative paper, analyze the roles you play in your family. Some of these may be typical cultural roles—such as father, mother, daughter, son, or oldest child—and some may be more diffuse roles—such as "clown," "teacher," "martyr," or "people pleaser."

PRACTICE TEST

The following items will help you evaluate your understanding of this lesson. Use the answer key at the end of the lesson to check your answers or to locate material related to each question.

Multiple-Choice

Select the one choice that best answers the question.

1. One of the main objectives of the textbook is to help you apply your sociological imagination to an understanding of
   A. mental processes such as perception, sensation, and emotion.
   B. the psychological dimension of life.
   C. the relationship between physical and biological systems.
   D. what social forces are shaping your biography and those of people you care about.

2. Which of the following is NOT an aspect of the social environment that sociologists might study?
   A. Those parts of the brain controlling sensation
   B. Religious behavior
   C. Conduct in the military
   D. The activities of voluntary associations like parent-teacher groups and political parties

3. A sociologist studying the migration of Asians to North and South America would be analyzing social reality at what level?
   A. Macro
   B. Middle
   C. Micro
   D. Mid-range

4. Which of the following is NOT true of social interaction?
   A. It is symbolic.
   B. It involves the exchange of messages that have meaning.
   C. It allows people to define each other as they interact.
   D. It can occur through an individual's dialog with himself or herself.

5. When Roberta defined Greg as a man, she was showing the importance of
   A. culture in the interaction process.
   B. her own ego in the interaction process.
   C. biological characteristics in the interaction process.
   D. instinct in the interaction process.

6. Socialization is defined as the ways in which people
   A. change social institutions through cooperation and competition.
   B. learn to conform to their society's norms, values, and roles.
   C. meet each other in social situations.
   D. converse freely as they form social relationships.

7. According to the Telecourse guide, children who receive messages that label them as "bad" or "less than" may grow up
   A. without intellectual ability.
   B. at a slower rate than other primates.
   C. with feelings of inferiority.
   D. too fast.

8. Culture includes all the following EXCEPT
   A. genetic inheritance.
   B. modes of thought.
   C. patterns of production.
   D. writing.

9. Which of the following is NOT involved in the way culture is passed down from one generation to the next?
   A. Communicative interaction
   B. Speech
   C. Gestures
   D. Genetic transmission

10. According to the Telecourse guide, which of the following is NOT true of personal social relationships?
    A. They are time-consuming.
    B. They are maintained only with effort.
    C. They are symmetrical.
    D. They are based on regular interaction.

11. The web of relationships that people form as they interact is referred to as social
    A. groups.
    B. category.
    C. structure.
    D. community.

12. Which of the following is an example of a social group?
    A. Riders on a bus
    B. Crowd on the street
    C. Married couple
    D. People with blonde hair

13. Sets of primary and secondary groups in which the individual carries out important life functions, such as raising a family, earning a living, and finding shelter, are called
    A. social structures.
    B. societies.
    C. social groups.
    D. communities.

14. Those who make up the legal community throughout the world provide an example of a
    A. society.
    B. non-territorial community.
    C. territorial community.
    D. social category.

15. Which of the following is an example of a society as defined in the Telecourse guide?
    A. Society for the Prevention of Cruelty to Animals
    B. A family
    C. Canada
    D. People living between the Rio Grande and the Arctic Circle

16. Sociology is the science of the
    A. human psyche.
    B. physical or natural world.
    C. mental processes of humans.
    D. methods people use to interact.

17. In which of the following is there predictability about how a person will behave in different circumstances?
    A. A social relationship
    B. A social category, such as gender
    C. A mob
    D. A crowd

18. What type of group is the family?
    A. Secondary
    B. Primary
    C. Reference
    D. Collectivity

19. Which of the following is considered by sociologists to be a way that their field can help individuals to understand their own lives?
    A. By becoming aware of our emotions, we can get along better with others.
    B. By studying other societies, we can discover how to better compete with them.
    C. By understanding how race relations are shaped by social structure, we can break down racial boundaries.
    D. By examining procedures used to formulate population figures, we can formulate accurate statistical conclusions.

ANSWER KEY

The following provides the answers and references for the practice test questions. Objectives are referenced using the following abbreviations: T = text, R = Telecourse Guide Reading, and V = Video.

Answers	Lesson Goals	Objectives	References
1. D	1	T1	Kornblum, p. 5
2. A	1	T2	Kornblum, p. 5
3. A	1	T3	Kornblum, pp. 6-7
4. D	2	R#1-1	TG Reading #1
5. A	2	R#1-2	TG Reading #1
6. B	2	R#1-3	TG Reading #1
7. C	2	R#1-4	TG Reading #1
8. A	2	R#1-5	TG Reading #1
9. D	2	R#1-6	TG Reading #1
10. C	3	R#2-1	TG Reading #2
11. C	3	R#2-2	TG Reading #2
12. C	3	R#2-3	TG Reading #2
13. D	3	R#2-4	TG Reading #2
14. B	3	R#2-4	TG Reading #2
15. C	3	R#2-5	TG Reading #2
16. D	4	V1	Video
17. A	4	V2	Video
18. B	4	V3	Video
19. C	4	V4	Video

Notes and Assignments:

Lesson 2

Social Interaction, Conflict, and Change

LESSON ASSIGNMENT

Review the following assignment in order to schedule your time appropriately. Pay careful attention; the titles and numbers of the textbook chapter, the telecourse guide, and the video program may be different from one another.

Text:

Kornblum, *Sociology in a Changing World*,
Chapter 1, "Sociology: An Introduction," pp. 18-27.

Reading #1:

Telecourse Guide for The Sociological Imagination, pp. 21-22.

Reading #2:

Telecourse Guide for The Sociological Imagination, pp. 23-25.

Video:

"Social Interaction, Conflict, and Change,"
from the series, *The Sociological Imagination.*

OVERVIEW

It was early December in 1955, the day was ending, and it was already dark outside. A woman, tired from her long day of work, dropped down into the first empty seat she found on the bus. After a few stops, the bus began to fill. The bus driver then ordered the woman and her fellow passengers with dark skin to move to the back of the bus, so the people with white skin, could have seats.

The woman was so tired that she surprised herself: She refused to rise. In fact, she did not get up until the police came and took her to jail.

That same Thursday evening, the word spread quickly throughout the black community. A bus boycott would begin in response to the arrest of Rosa Parks.

Just four days later, fifty people—ministers and civic leaders—met in the Holt Street Baptist Church. An articulate young black minister was elected the first president of the Montgomery Improvement Association (MIA), which would help coordinate car pools, low fares in black-owned cabs, and other means of transportation for 17,000 riders a day.

During the next year, bolstered by heavy press coverage and more than $250,000 in donations from all over the world, the MIA expanded to support a staff of ten. From this grew what is now known as the civil rights movement. Martin Luther King, Jr., chosen as MIA president that Monday night in December 1955, would become a passionate and effective leader of the social movement that would unalterably change American culture and social structure.

This major social change was sparked by a single human interaction, a conflict between a white bus driver and an African-American woman tired from a hard day's work—and tired of a system of prejudice and discrimination that oppressed her and her people.

Two days after Rosa Parks' conviction, Americans would remember another conflict, begun on December 7, 1941, when the Japanese sunk or damaged nineteen ships and killed 2,300 people at Pearl Harbor. The ensuing war, with is massive human toll and its forced infusion of women into the work force, also permanently changed the social and cultural landscape of the United States.

In the previous lesson, you were introduced to the idea that social interactions lead to social structures. In the web or system of relationship that constitute these structures, we can see a major theme of sociological literature: social order. In all of social life there is some degree of order.

Social organization—a term frequently used by sociologists—refers to social groups of all kinds, both formal and informal. But social organization also implies a coordination among social relationships: the cooperation between people that is necessary for any social system to persist.

For example, the meeting at Holt Street Baptist Church, the hundreds of telephone calls, and the many meetings of church congregations across the country were all to *coordinate* or *organize* actions of protest and indignation. Of course, this cooperation was aimed at altering another social order: the racist one that held sway throughout the nation at the time. You will see this theme of changing the social order recur throughout this course.

Although social order and social systems constitute a major theme in sociology, in this lesson you encounter yet another perspective. You see how social interaction and conflict lead to *social change*. One of the enduring contributions of sociology is its analysis and understanding of processes of social conflict and social change, two realities that often perplex and disturb us.

You also discover the way sociologists analyze different levels of social life:

1. For instance, the conflict between Rosa Parks and the bus driver was social behavior at a *micro level.*

2. Late in 1956, the U.S. federal courts ruled Alabama's bus-segregation laws unconstitutional. The legal fight waged in this governmental institution is an example of a conflict at the *middle level* of social behavior.

3. Those court decisions signaled the long process of social change that would occur as a result of the civil rights movement. The changes in attitudes, governmental policies, and social relationships that have taken place on a broad societal level are known as changes at the *macro level.*

This lesson explores interaction, conflict, and change at all three of these levels. For example, the video program examines how change affects individuals (micro level) and how it has occurred in an institution of higher education (middle level). But it also looks at social change in a community and in a developing nation (macro level).

Sociologists have always studied social change. The assumption of your textbook and of this course is that social change not only is a constant in our lives and in societies throughout the world but is a major driving force within the discipline of sociology itself. It is our hope that learning more about this layer of social reality will help you to function better in your social groups, in your community, and in a complex and ever-changing society.

LESSON GOALS

Upon completing this lesson, you should be able to:

1. Analyze the major sociological perspectives.
2. Analyze social change at the community level, and analyze how social change produces role conflict.
3. Analyze how social stratification, prejudice and discrimination, social power, and collective behavior are implicated in social change.
4. Analyze the different levels of social conflict and social change, and analyze how they affect individuals.

TEXTBOOK OBJECTIVES

The following textbook objectives are designed to help you get the most from the text. Review them, then read the assignment. You may want to write notes to reinforce what you have learned.

Text: Kornblum, *The Sociological Imagination*, Chapter 1.

1. Describe interactionism, including rational choice theory and symbolic interactionism.
2. Explain the functionalist perspective.
3. Explain conflict theory and the contributions of Karl Marx to the conflict perspective. Identify the important questions for sociologists who study conflict and power.

READING #1
FROM THE MICRO TO THE MACRO

READING OBJECTIVES

The following reading objectives are designed to help you get the most from Reading #1. Review them, then read the assignment. You may want to write notes to reinforce what you have learned.

Reading #1: *Telecourse Guide for The Sociological Imagination,* pp. 21-22.

1. Describe the difference between gemeinschaft and gesellschaft. Include social changes that were involved in the formation of each.
2. Differentiate between primary and secondary groups.
3. Explain the meaning of role conflict. Identify what social conditions cause it. Be able to recognize examples of role conflict.

When America Was Young

When America was young, before the United States was even formed, its inhabitants lived in an agricultural society. Most people were farmers. Life was simple. Familiarity and a feeling of kinship pervaded social relationships, making the community seem like a big family. This type of social structure is known as a *gemeinschaft*, meaning the close, personal relationships of small groups and communities.

Another related term used to describe the groups within communities is primary groups. Most of the groups in agrarian America were *primary groups* that is, face-to-face, intimate units in which people were viewed as total persons, rather than seen primarily in terms of their specialized roles.

As society becomes more complex and industrialized, people do things other than farming. Most live in cities and work as carpenters, plumbers, managers, or textile workers. Thousands of people, filling different specialties, mass in small geographic areas. Complex, industrialized societies develop well-organized social structures in which there are many impersonal relationships.

These societies and the cities within them are *gesellschaft* social structures. Factories and office bureaucracies dominate day-to-day life in the modern, gesellschaft society.

Although people in the *gesellschaft* still have many primary groups, they have to learn to adapt to the larger, more complex and more formal groups that overshadow the cities. In these secondary groups, relationships are more impersonal, often based on contracts, with people interacting with each other in segmented roles.

There are many conflicts that accompanied the change from *gemeinschaft* and its primary groups to *gesellschaft* and its secondary groups. One such clash that sociologists observe is referred to as role conflict. It occurs when, in order to perform one role, a person must violate another important role. Parents who are also employees may experience this kind of conflict when their supervisors ask them to put in extra time, which cuts into the time with their children. The role in the secondary group is not compatible with the role in the primary group.

By now, you are beginning to see that our individual lives are not lived in a vacuum. Rosa parks was minding her own business that December evening as she rested her tired body by taking the bus home from work, but social forces beyond her intervened in her life.

Suddenly, her "micro world" came head-to-head with the racist proscriptions of the "macro world" of her community and society.

Obviously, as social structures changed with industrialization, people's individual lives also changed. They had new conflicts and challenges to face. In modern societies, people must continually try to reconcile the demands of their macro social world (for example, economy and work) with the needs of their micro social world (family and other primary groups).

READING #2
SOCIAL INEQUALITIES, COLLECTIVE BEHAVIOR, AND SOCIAL CHANGE

READING OBJECTIVES

The following reading objectives are designed to help you get the most from Reading #2. Review them, then read the assignment. You may want to write notes to reinforce what you have learned.

Reading #2: *Telecourse Guide for The Sociological Imagination,* pp. 23-25..

1. Explain the meaning of social stratification.
2. Describe the difference between prejudice and discrimination.
3. Recognize examples of stereotypes.
4. Describe and give examples of collective behavior.
5. Explain the meaning, and give examples, of both social power and social change.

Conflict and Inequality

Karl Marx said that the history of the world is a history of class struggle. Marx recognized the importance of social conflict in social life. Since then, sociologists have examined various levels of conflict. In the previous segment you saw how social changes and structural factors can cause conflict in our lives as individuals—at the micro level. This segment focuses primarily on conflict at the macro level and examines some of the central sociological concepts related to social conflict and social change.

Many of the stories you read in the newspaper and hear on television and radio are a result of conflicts between the "haves" and the "have-nots." This is because the social structures of modern societies are stratified. Some segments of society have more of the socially valued rewards than others. *Social stratification* is a society's system for ranking people hierarchically (i.e., from high to low) according to certain attributes such as income, wealth, power, prestige, age, sex, ethnicity, gender, and religion.

Even though Americans generally believe in equal opportunity for all, the fact is that money, power, prestige, and resources are not equally distributed in this society. The terms "upper class," "middle class," and "lower class" describe the class structure that is the main manifestation of social inequality or social stratification.

Inequality and conflict in modern societies are often based on prejudice and discrimination. *Prejudice* is an attitude. It involves prejudging a person on the basis of some real or imagined characteristics (stereotypes) of a group of which that person is a member.

One expression of prejudice is the formation of stereotypes. *Stereotypes* are inflexible images of a racial or cultural group that are held without regard to whether or not they are true. Examples of these over-generalized images are: Jews as cheap, blacks as lazy and shiftless, Hispanics as violent and hot-tempered, women as emotional.

Prejudice often is used to justify treating people differently, as somehow subordinate or having less value. *Discrimination* is not an attitude, it is an action or series of actions. It refers to actual unfair treatment of people on the basis of their group membership. An example is not hiring people based on their ethnicity, age, or gender.

When this happens in "the land of opportunity," it creates a social problem. Rosa Parks suffered the pains of prejudice and discrimination during those fateful days of December 1955.

The civil rights movement and other social movements are examples of collective behavior, which involves people acting differently when they are part of a group than they would act by themselves. Collective behavior refers to spontaneous actions of people who react to situations that they perceive as uncertain, threatening, or extremely attractive. Some examples are riots, mobs, and crazes. Much social change is accompanied, or created by collective behavior.

Most social conflicts we have spoken of so far also have something to do with social power. *Power* is the probability that one actor within a social relationship will be in a position to carry out his or her own will despite resistance. When a boss fires an employee even though the worker strongly objects, the boss has exercised power. If you are part of the working class, you have less power in our social-stratification system than a person in the upper-middle class. If you are considered part of the middle class, you have less power than a person in the upper class. In most cases, if you have experienced *systematic* prejudice and discrimination, it is probably because you have less power than your offenders.

The civil rights movement and other reform movements often are attempts to achieve more social power for the aggrieved parties, so you can see the importance of the concept of social power in society, and therefore in sociology.

We have used the term "social change" several times in this lesson. Just what does it mean? Does it refer to individual life changes we go through in our personal lives? Does it mean changes in groups or formal organizations that we read about in the newspapers? Actually, while social change can affect both of these, sociologists have a specific and special meaning for the term that goes well beyond these more limited changes. *Social change* refers to variations over time in populations and communities, in patterns of roles and social interactions. The societal changes from agrarian and *gemeinschaft* to industrial and *gesellschaft* social structures are examples.

In conclusion, many social changes take place as a result of conflicts over inequalities. Some of those inequalities are due to social stratification and some of them are based on prejudice and discrimination. Social movements are attempts to achieve power and to "move" society, that is, to bring about a social change.

VIDEO OBJECTIVES

The following video objectives are designed to help you get the most from the video segment of this lesson. Review them, then watch the video. You may want to write notes to reinforce what you have learned.

Video: "Social Interaction, Conflict, and Change"

1. Describe the consequence of human interaction in conflict. Determine the causes of social change.

2. Describe the conflicts and the accompanying social changes taking place in Dallas and nearby towns that are typical of those occurring throughout the United States between large cities and the small rural towns near them.

3. Describe the conflicts and resultant changes occurring between nations, using conflicts and social changes accompanying the independence of Third World nations, such as Trinidad and Tobago, as an example.

RELATED ACTIVITIES

These activities may be used by your instructor as written assignments or as discussion topics. They may also be included as essay questions on your tests.

1. With which of the three major sociological perspectives do you identify the most, that is, which is most similar to your view of society and social life? Explain what life experiences have led you to hold your view.

2. Describe an example of a role conflict that you have experienced.

3. Write a short paper comparing your socially valued rewards with those of someone you know who is in a higher or lower social class than you. Describe some of the external signs of this difference, such as clothing, car, entertainment, etc.

4. Describe an incident in which you personally experienced prejudice or stereotyping. What was the prejudice you experienced or the stereotype applied to you? Describe the situation, the people involved, and your reaction in a short narrative.

5. Write a brief description of a situation in which you exercised power over someone else. You may use work, family, an organization, or any other social situation as your example.

PRACTICE TEST

The following items will help you evaluate your understanding of this lesson. Use the answer key at the end of the lesson to check your answers or to locate material related to each question.

Multiple-Choice

Select the one choice that best answers the question.

1. The sociological perspective that views social order and social change as resulting from all the immense variety of repeated interactions among individuals and groups is
   A. interactionism.
   B. functionalism.
   C. conflict theory.
   D. Marxist theory.

2. Which of the following perspectives ask how society manages to carry out the functions it must perform in order to maintain social order, feed large masses of people each day, defend itself against attackers, and produce the next generation?
   A. Symbolic interactionism
   B. Rational choice theory
   C. Functionalism
   D. Conflict theory

3. Karl Marx believed that division of people in a society into different classes, defined by how they make a living, always produces
   A. conflict.
   B. exchange.
   C. profit.
   D. equilibrium.

4. A society whose social structures are marked by close personal relationships found in small groups is called
   A. *gesellschaft*.
   B. associational.
   C. *gemeinschaft*.
   D. complex.

5. A group in which relationships are impersonal, often based on contracts, is known as a
   A. primary group.
   B. secondary group.
   C. *gemeinschaft*.
   D. small group.

6. When employees who are also parents are asked by their supervisors to work extra hours, the employees may experience
   A. role transference.
   B. master-status conflict.
   C. social change.
   D. role conflict.

7. A system for ranking people hierarchically according to various attributes, such as income, wealth, power, prestige, age, sex, ethnicity, gender, and religion, is called social
   A. ranking.
   B. dominance.
   C. inconsistency.
   D. stratification.

8. Doc owns a liquor store and will not allow Italians to enter because other ethnic groups have threatened him with a boycott if he does. Doc is married to and very much in love with an Italian woman and loves her family.

   Doc's behavior toward his would-be customers is an example of
   A. prejudice.
   B. a bad attitude.
   C. discrimination.
   D. selective perception.

9. "Spontaneous actions of people who react to situations that they perceive as uncertain, threatening, or extremely attractive" is a description of one aspect of
   A. collective behavior.
   B. impulse syndrome.
   C. involuntary behavior.
   D. automatic response.

10. Power is the probability that one actor within a social relationship will be in a position to carry out his or her own will in
    A. cooperation with others.
    B. spite of resistance.
    C. bureaucracies or other formal organizations.
    D. roles of leadership.

11. The most important agent of social change is
    A. conflict.
    B. cooperation.
    C. religion.
    D. culture.

12. Which of the following conflicts was brought about by the urban revolution?
    A. Habits and norms of heterogeneous populations living in the same area create conflict.
    B. Technological development taking place within cities brings about conflict between the upper and upper-middle classes.
    C. Cities that are dominated by one religious group often experience conflict caused by aggrieved minority religions.
    D. Increasing stratification within a geographic area causes conflict among the social classes.

13. Which of the following conflicts between nations has caused social change?
    A. Conflicts between radical liberals in other countries and conservatives in America
    B. Conflicts between powerful colonizing nations and indigenous peoples
    C. Conflicts between authorities and social deviants emigrating from other nations
    D. Conflicts between businesses in various nations and providers of scarce resources

ANSWER KEY

The following provides the answers and references for the practice test questions. Objectives are referenced using the following abbreviations: T = text, R = Telecourse Guide Reading, and V = Video.

Answers	Lesson Goals	Objectives	References
1. A	1	T1	Kornblum, p. 18
2. C	1	T2	Kornblum, p. 19
3. A	1	T3	Kornblum, p. 20
4. C	2	R#1-1	TG Reading #1
5. B	2	R#1-2	TG Reading #1
6. D	2	R#1-3	TG Reading #1
7. D	3	R#2-1	TG Reading #2
8. C	3	R#2-2-3	TG Reading #2
9. A	3	R#2-4	TG Reading #2
10. B	3	R#2-5	TG Reading #2
11. A	4	V1	Video
12. A	4	V2	Video
13. B	4	V3	Video

Notes and Assignments:

Lesson 3

Sociological Thinking and Research

LESSON ASSIGNMENT

Review the following assignment in order to schedule your time appropriately. Pay careful attention; the titles and numbers of the textbook chapter, the telecourse guide, and the video program may be different from one another.

Text:

Kornblum, *Sociology in a Changing World*,
Chapter 2, "The Tools of Sociology," pp. 28-57.

Reading:

Telecourse Guide for The Sociological Imagination, pp. 36-39

Video:

"Sociological Thinking and Research,"
from the series, *The Sociological Imagination.*

OVERVIEW

Sharon Hicks-Bartlett was brought up in a Chicago suburb. Her mother warned her of the possible harm that could be done by the social workers who were in their community to "help" the poor black people there. But young Sharon was curious about why she and her people would need help and special treatment. The neighborhood was basically poor, sure, but it included a prominent minister and a lawyer and other professionals, and a physician lived next door. Who were these social workers, anyway, and what were they there for?

Even as a child, Sharon had the germ of a sociological imagination growing in her mind. This curiosity kept her mind open to many possibilities for her own future.

Later, as an adult on her way to attend college classes in the city, a more mature Sharon had to drive through another, even more impoverished, and geographically and economically isolated, African-American community. Her discomfort with the conditions there haunted her, spawning more questions. Was there something about the social life and history of this area that caused the poverty to persist, to be so uniform and widespread?

At the university, the ever-curious Sharon first studied history, but her interest in families and people soon thrust her into the field of gerontology. Eventually, those haunting questions from her childhood and her daily travel inevitably led her to sociology.

Ms. Hicks-Bartlett spent a whole year reading the old ethnographies—community studies—done by early researchers. They sparked her imagination, already driven by that virulent curiosity. She swallowed her fears and, through patience and determination, gained entree into, and became accepted as a researcher by the people in that poor community she had driven through every day.

In her writings she calls the community "Meadowview" to protect the anonymity of the people there. For more than half a decade she has been doing sociological research at its funerals and weddings, talking to drunks—the defeated ones—and to the old people—the repository of the community's history—forming relationships in all kinds of groups in many different settings in Meadowview.

After hours of conversations with her "informants," her research subjects, she recorded her field notes about what she'd heard and seen, and gradually patterns began to emerge about the way poverty had been reproduced in Meadowview.

This is a story of just one sociologist and her research. But, exactly as happened with Sharon Hicks-Bartlett, all research starts with an idea, an experience, a feeling—and the curiosity to ask: Why? This lesson uncovers what happens next in the research process.

Although every sociological study is different, all researchers follow certain basic steps or procedures—the same ones followed by Galileo, Madame Curie, Jonas Salk, and W.E.B. DuBois. You may never be a scientist, but you *can* appreciate science and be come more scientific in your thinking.

As our look at sociological research illustrates, science is not just for experts in white lab coats. It is for all who have a curiosity and yearning to gain accurate information about their physical and social worlds.

This lesson introduces you to science as sociologists practice it. As you study the lesson, continue to keep your eyes open to the special subjects and perspectives that sociologists take into their investigations of life in any society.

LESSON GOALS

Upon completing this lesson, you should be able to:

1. Analyze the steps in the scientific method and the way social scientists try to gather information about the social world.
2. Analyze the methods used by sociologists.
3. Analyze the qualities of science and how sociologists interpret their data.
4. Analyze how sociologists view the world and its people, and determine why the scientific method plays such an important role in obtaining valid information.

TEXTBOOK OBJECTIVES

The following textbook objectives are designed to help you get the most from the text. Review them, then read the assignment. You may want to write notes to reinforce what you have learned.

Text: Kornblum, *Sociology in a Changing World*, Chapter 2.

1. List the steps in the scientific method.
2. Define the terms *empirical study, hypothesis,* and *variable.*
3. Differentiate between a dependent and an independent variable, and be able to recognize examples from studies of welfare dependency.
4. List the questions and interest areas of the ecological, functionalist, and conflict perspectives that one is likely to see reflected in sociological literature. Describe this, using suicide as an example.
5. Explain and give examples of the different types of observation used in a sociological research.

6. Explain and exemplify the different types of experiments used in social research. Explain the significance of the "Hawthorne Effect."

7. Explain why survey research is such a powerful tool for social researchers. Describe the different types of surveys sociologists use. Recognize examples of sample surveys.

READING
OTHER SOCIAL SCIENCE ISSUES

READING OBJECTIVES

The following reading objectives are designed to help you get the most from the Reading. Review them, then read the assignment. You may want to write notes to reinforce what you have learned.

Reading: *Telecourse Guide for The Sociological Imagination*, pp. 36-39.

1. Describe what is involved in the collection and analysis of scientific research data. Use the research of Sharon Hicks-Bartlett as an example.

2. Explain what is meant by the qualities of validity, reliability, and objectivity in scientific research.

From Sweat to Intellect: Collecting and Interpreting the Data

The story of the life and research of Sharon Hicks-Bartlett presented in the Overview of this lesson touches on the several steps that scientists go through as they try to make sense of social life.

Sociological researchers begin with the problems, questions, or issues that drive or motivate them. Then they examine existing sociological work on those questions. Next, the researchers move to a more detailed plan about how the research will actually be carried out. The textbook describes these steps.

However, one stage of the process not specifically enumerated there is the data-collection phase. Some students of social science have called it the "sweat stage." This phase involves the collection of the information that will become the raw material of the research project.

The type of study Sharon Hicks-Bartlett did is known in social science as ethnographic research. An *ethnography* is a study in which the social researcher observes people in their daily lives and routines over an extended period of time. The days and nights Hicks-Bartlett spent talking to and observing the people in Meadowview, and the tedious process of recording those observations, comprise her data-collection stage. The energy and time involved in this process can be immense and wearisome.

Collecting data—whether in interviews, questionnaires, use of secondary sources, or observation—often is drab, monotonous, and laborious. But it can also be stimulating and energizing. Hicks-Bartlett's experience suggests that researchers should find research projects that suit their individual personalities and interests and about which they are intellectually curious. That way they can more easily survive the rigors of data collection.

Eventually, Hicks-Bartlett began to see patterns emerging in the stacks of field notes she had written on her observations in Meadowview. She began to understand that the community was reproducing—through the myriad of social interactions and informal teachings—the very things it was trying to break out of. She became aware of the entanglement of family and community. Meadowview's culture and its social institutions and structures, were interacting to produce the poverty, and the economic and social isolation, so troublesome to this community. But these interpretations emerged over the years only as Sharon Hicks-Bartlett began to analyze the data she had collected.

Data analysis takes some expertise; it is not mastered by a neophyte or someone unfamiliar with the data. *Analysis means that the researcher discerns patterns—patterns of behavior, social relationships, groups, and social arrangements.* Studies using quantitative methods may rely on computers for much of the analysis, but qualitative methods require a different kind of interpretation of data.

For example, Hicks-Bartlett discovered that Meadowview had evolved a sort of "kin" system into which it was extremely difficult to be allowed entrance. Only after she had become very familiar to, and thus trusted by, the people of Meadowview was she allowed to freely observe their family arguments, gossip, and informal chatter. The discovery of this kinlike system was a fruit of her analysis stage of research.

Qualities of Science

The rules or steps in the scientific method are general guidelines for social research. They are not followed in a lock-step manner by people actually doing sociological studies. The purpose of these rules is to provide systematic direction for the scientist. This is because what is important for good science is that it evidences the qualities of *validity*, *reliability*, and *objectivity.*

For example, if other researchers entered the Meadowview community of Sharon Hicks-Bartlett's study and proceeded to do a study of parents' attitudes toward child rearing by administering a questionnaire, would they obtain the information they really wanted? Their results might not be *valid*, due to their respondents' suspicion of outsiders. *Validity is the extent to which scientists are measuring what they intend to measure.*

If you own a car, you want it to start every morning when you turn on the ignition. Consumers rate cars on their reliability. In scientific research, *reliability means that repeated research studies on a particular group or process will yield results consistent with the original study.* Sharon Hicks-Bartlett's study will be found to be reliable if she or other researchers study the same or similar communities and end with the same or similar conclusions. This openness of the research process to scrutiny by others is one of the distinctive qualities of science.

Has anyone ever told you to try to be objective? What that usually means is that you should not become so emotionally involved. *Objectivity is an attitude toward a situation in which personal interest and cultural and group biases are controlled.* It means that scientists must maintain their neutrality as they proceed through the steps in the scientific method.

Complete objectivity is probably impossible. Because all scientists have some personal interest in what they are studying, and because they must interpret their findings, there are moments when they become subjective, rather than objective. In the physical sciences, such as astronomy or chemistry, neutrality is more feasible than in the social sciences, since in the physical sciences the researcher isn't part of the subject matter under study. The biologist studying cell growth under a microscope is, of course, not there on the slide beneath the lens with the subject cells. In sociological research, however, the sociologist is often part of the society or culture under study.

Sociologists, like all other scientists, try to be as objective as possible. But it is especially important for social researchers to be aware of their own cultural and personal biases, and to state them along with their findings, so that others can intelligently evaluate the research.

Sociology, then, is a respected science. And *science is a systematic process of gathering accurate and reliable knowledge about the physical and social world.* The qualities of science that all researchers and theorists try to achieve as they observe the world are validity, reliability, and objectivity.

VIDEO OBJECTIVES

The following video objectives are designed to help you get the most from the video segment of this lesson. Review them, then watch the video. You may want to write notes to reinforce what you have learned.

Video: " Sociological Thinking and Research"

1. Discuss how reliable common sense and casual observation are in understanding social life. Explain how these differ from a social-scientific perspective.
2. Describe the research project featured on the video program as it progresses through each step in the scientific method. Be able to match examples of the research with each of the steps of the scientific method as explained in the video program.
3. Explain how the research on Times Square has been used. Determine the hypothesis of Kornblum and Williams' research on Times Square.

RELATED ACTIVITIES

These activities may be used by your instructor as written assignments or as discussion topics. They may also be included as essay questions on your tests.

1. If you had a large grant to do a sociological study, what subject, issue, or problem would you be interested in studying? Why did you choose that

topic; that is, what was there about your experiences, background, and interests that caused you to focus on that topic?

2. Your neighborhood offers many opportunities to do sociological research using unobtrusive measures. Describe at least two possible studies that could be done in your neighborhood using this method. How would you gather your evidence? What is it about people's lifestyles, groups, or social relationships that you could discover using this technique?

3. Design a questionnaire to discover student attitudes toward telecourses. Compose at least four questions to obtain the information you need. You may use "open questions" or "closed questions." Describe the kind of sample you would use, as well as how you would identify your respondents and acquire the responses to the questionnaire, that is, how you would gather your data.

4. After viewing the video program, describe your reaction to, and evaluation of, the research done by Professors William Kornblum and Terry Williams.

5. What makes the information gained through sociological research different from information you read in everyday newspapers and magazines?

PRACTICE TEST

The following items will help you evaluate your understanding of this lesson. Use the answer key at the end of the lesson to check your answers or to locate material related to each question.

Multiple-Choice

Select the one choice that best answers the question.

1. Which of the following is NOT a step in the scientific method?
   A. Programming the computer
   B. Reviewing the literature
   C. Formulating the research questions
   D. Analyzing the data

2. Characteristics of individuals, groups, or entire societies that can vary from one case to another are called
   A. variances.
   B. hypotheses.
   C. samples.
   D. Variables.

3. Emile Durkheim's study of suicide was a way of exploring the larger concept of
   A. psychological trauma among certain populations.
   B. personality dysfunctions of victims of suicide.
   C. historical development of mental hospitals.
   D. integration or the lack of it in society.

4. Suicide rates are higher for unmarried people than for married people.

   In the above hypothesis, the dependent variable is
   A. marriage rate.
   B. age of married people.
   C. suicide rate.
   D. number of children.

5. From which sociological perspective would a researcher be likely to ask: "What groups or organizations are involved?"
   A. Conflict
   B. Interactionist
   C. Functionalist
   D. Ecological

6. When the sociologist becomes involved in the daily life of the people whom he or she is observing, what research method is being used?
   A. Unobtrusive measures
   B. Participant observation
   C. Visual sociology
   D. Experiment

7. Solomon Asch's research in which two groups were required to match correct lines in a figure is an example of
   A. unobtrusive measures.
   B. participant observation.
   C. controlled experiment.
   D. survey research.

8. The most ambitious and most heavily used sociological surveys are the
   A. national censuses.
   B. CBS News & New York Times polls.
   C. Literary Digest and other similar polls.
   D. Shere Hite surveys.

9. The collection of data in sociological research is the
   A. stage at which researchers draw their conclusions about the research project.
   B. means of obtaining information that will become the raw material of the research project.
   C. point at which the researcher has no control over the research design.
   D. first step in the scientific method.

10. The extent to which scientists are measuring what they intend to measure is known as
    A. objectivity.
    B. validity.
    C. reliability.
    D. Precision.

11. Which of the following is NOT true of sociology?
    A. It provides an accurate way of studying social interaction.
    B. It represents a systematic and scientific approach to social reality.
    C. It generates more scientific results than psychology does.
    D. It often contradicts common-sense ideas.

12. A hypothesis of Professor William Kornblum and Professor Terry Williams's research project on Times Square, described in the video program on sociological thinking and research, was that
    A. prostitution is caused by child abuse in the family of origin.
    B. movie going is directly related to social class.
    C. government officials are more likely to take bribes if they have been in office for a very long time.
    D. pornography, if eliminated in the Forty Second Street area, will reappear elsewhere.

13. According to Professor William Kornblum, the research on the Times Square area featured on the video program on sociological thinking and research has been used by
    A. religious groups to recruit possible members.
    B. mass media to target audiences for local news programs and other local programming.
    C. private organizations that are able to purchase the complex data and data analysis.
    D. participants in legal battles concerning the gathering of parcels of property in the area.

ANSWER KEY

The following provides the answers and references for the practice test questions. Objectives are referenced using the following abbreviations: T = text, R = Telecourse Guide Reading, and V = Video.

Answers	Lesson Goals	Objectives	References
1. A	1	T1	Kornblum, p. 30
2. D	1	T2	Kornblum, p. 31
3. D	1	T3	Kornblum, p. 31
4. C	1	T4	Kornblum, p. 31
5. C	1	T4	Kornblum, p. 34
6. B	2	T5	Kornblum, p. 35-36
7. C	2	T6	Kornblum, p. 37-40
8. A	2	T7	Kornblum, p. 41
9. B	3	R1	TG Reading
10. B	3	R2	TG Reading
11. C	4	V1	Video
12. D	4	V2	Video
13. D	4	V3	Video

Notes and Assignments:

Lesson 4

Culture

LESSON ASSIGNMENT

Review the following assignment in order to schedule your time appropriately. Pay careful attention; the titles and numbers of the textbook chapter, the telecourse guide, and the video program may be different from one another.

Text:

Kornblum, *Sociology in a Changing World*,
Chapter 3, "Culture," pp. 58-89

Reading:

There is no Reading for this lesson.

Video:

"Culture,"
from the series, *The Sociological Imagination*.

OVERVIEW

Maria held the ancient rosary in her hands, allowing the tiny wooden beads to fall gently between her fingers. The tarnished chain, nearly worn apart at the "Our Father" beads, betrayed its age. She glanced over at the painting of her great grandmother, the original owner of the beads, and wondered how many hours her great grandmother must have spent on her callused knees, praying beside the bed of a sick child or on the dirt floor of the cathedral in her beloved Mexico.

Although Maria didn't have the faith of her forebears, she cherished the worn rosary. This object of material culture symbolized her cultural heritage: the beliefs, the values, the morals and folkways that guided and comforted, and made life bearable and meaningful for, generations before her.

Your family too may have some objects—photos, antiques, medals—revered for the cultural history and tradition they embody. Social scientists are intrigued by these artifacts, partially because they give us hints about the way of life of those who produced them. But sociologists are especially interested in how those ways still persist in the minds of today's culture bearers. What are the "menti-facts" behind the artifacts? That is, what are the belief and normative systems, the ideas and values, that cause people to produce the things that are known as artifacts and technology: chairs, plows, cars, powder, cameras, posters, coolers, and petroleum?

Like Maria, perhaps you've been curious about your own cultural heritage. Or maybe you've been mystified or even frustrated by the actions and words of the generation before you! It's this kind of intellectual or emotional discomfort that will motivate you—as it motivates many sociologists—to sharpen your sociological imagination in this lesson about culture.

In this lesson you find some sociological answers to these questions. You are challenged to consider your own cultural ideas and biases in a new light. You discover the importance of language and see how diverse cultures affect each other.

With this new awareness you will be better equipped to participate in and mold the multicultural society and the increasingly interdependent world in which we all live.

LESSON GOALS

Upon completing this lesson, you should be able to:

1. Analyze culture and the normative aspects of culture in terms of how they affect you and those around you.

2. Analyze how cultures develop and the role of language in culture, in order to acquire a critical understanding of social Darwinism.

3. Analyze diverse cultures, as well as how cultures affect each other.

4. Analyze the historical and environmental context of culture, and analyze how different subcultures in the United States address human needs in different ways.

TEXTBOOK OBJECTIVES

The following textbook objectives are designed to help you get the most from the text. Review them, then read the assignment. You may want to write notes to reinforce what you have learned.

Text: Kornblum, *Sociology in a Changing World*, Chapter 3.

1. Define culture and be able to recognize examples of the dimensions of culture.
2. Explain the meaning of norms. Recognize examples of norms.
3. Define and recognize examples of values.
4. Define laws.
5. Explain the meaning of mores. Recognize examples of mores.
6. Define folkways. Recognize examples of folkways.
7. Explain the main beliefs of the social Darwinists, especially Herbert Spencer and William Graham Sumner.
8. Describe the importance of language in relation to culture.
9. Explain the "linguistic relativity" hypothesis in its various forms.
10. Define ethnocentrism.
11. Define cultural relativity. Recognize non-textbook examples of cultural relativity.
12. Define acculturation and describe the two-way process of acculturation. Recognize examples of acculturation.
13. Define and cite examples of assimilation.
14. Define subculture. Recognize examples of subcultures.

VIDEO OBJECTIVES

The following video objectives are designed to help you get the most from the video segment of this lesson. Review them, then watch the video. You may want to write notes to reinforce what you have learned.

Video: "Culture"

1. Explain how culture is a response and an adaptation. Recognize examples of culture.

2. Recognize the two factors contributing to the preservation of a subculture.

3. Explain how these factors have played a role in the Cajun, Cherokee, and Mississippi Chinese subcultures.

4. Explain ethnocentrism. Describe and exemplify the cost of the white man's ethnocentrism for the Comanche.

RELATED ACTIVITIES

These activities may be used by your instructor as written assignments or as discussion topics. They may also be included as essay questions on your tests.

1. Although it is not accurate to say each of us has our own culture, each of us lives in a society, community, or family that bears a culture to us. In general terms, characterize your culture, that is, the culture with which you identify the most. In short, describe your "cultural heritage."

2. Describe three norms you accept and believe in. In your sentences, use normative words, such as should, must, ought. Example: One should say thank you when given a gift. Identify whether they are folkways, mores, or laws.

3. In single words, make a list of four of your values. Below your list, identify each of your words for these values by describing how each does and/or does not match traditional American values.

4. What kind of rewards and positive reinforcements were you given as a child by your significant others to help shape your behavior? What punishments

were used? Which of them were appropriate, and which were abusive, if any?

5. Give five examples of laws, customs, or norms that are part of your culture. What is the value represented by each? Describe how each provides a sense of stability or comfort in the face of changing conditions. For example: People on a submarine, on a ship, in a remote location, or in the military can constitute a culture of their own. What laws, customs, or norms are part of that culture?

6. List and describe words, dialect, or grammatical constructions that you use or have used that reflect your cultural heritage. Perhaps your family is from a social class, or ethnic, religious, or other subculture, that is different from the dominant culture. Even if you're not, each family has its own unique words or ways of saying things that may not match conventional language. Identify these language differences.

7. List and describe terminology that is used as part of your culture that others not in the same situation may not understand. Identify at least five of these language differences and explain what they mean. For example: People on a submarine, on a ship, in a remote location, or in the military can constitute a culture of their own. What terminology is used as a part of that culture?

8. Describe a situation in which you had difficulty communicating with someone because of language problems. Perhaps the other person spoke a language foreign to you or used a vocabulary with which you were unfamiliar.

9. Using your sociological imagination, what do you think individuals, groups, and societies can do to adopt an attitude of cultural relativity toward others who are culturally different? What can we do to diminish our own or a group's ethnocentrism?

10. Interview someone from another culture who has attempted to assimilate fully into the dominant culture, that is, to shake off any signs of being culturally distinct. Without judgment or criticism, ask questions to find out why the person attempted to assimilate. What experiences, persons, groups, or institutions influenced the individual to do so?

PRACTICE TEST

The following items will help you evaluate your understanding of this lesson. Use the answer key at the end of the lesson to check your answers or to locate material related to each question.

Multiple-Choice

Select the one choice that best answers the question.

1. Which of the following represents an element or an aspect of culture?
   A. Family meal
   B. Blade of grass on the Great Plains
   C. Sound of a swift stream
   D. Blinking of the eyes

2. You should study for several hours before a sociology exam.

   This is an example of
   A. values.
   B. sanctions.
   C. ideologies.
   D. norms.

3. Values are
   A. ideas that support or justify norms.
   B. specific rules of behavior.
   C. punishments for behavior that violates the norms.
   D. formalized norms of society.

4. Norms that are included in a society's officially written codes of behavior are called
   A. folkways.
   B. sanctions.
   C. values.
   D. laws.

5. All states prohibit murder.

   This prohibition of murder is an example of
   A. folkways.
   B. ideologies.
   C. Mores.
   D. sanctions.

6. Don't eat peas with your fingers.

   The above is an example of
   A. mores.
   B. institutionalized roles.
   C. laws.
   D. folkways.

7. Herbert Spencer believed that
   A. government should intervene in the lives of the powerless segments of society and stand by them in times of need.
   B. economy should be an arm of government and subject to political process and control.
   C. social evolution could not be improved by means of intentional action.
   D. social development and evolution can be brought about by effective public action through political institutions.

8. Sociologists who believed that laws to protect people less able to compete were doomed to failure because they ran counter to the force of social evolution are called
   A. Marxist sociologists.
   B. symbolic interactionists.
   C. social Darwinists.
   D. cultural relativists.

9. That language determines the possibilities for thought and action in any given culture is an expression of the
   A. theory of social Darwinism.
   B. linguistic-relativity hypothesis.
   C. cultural relativity analysis.
   D. theory of cognitive development.

10. The tendency to judge other cultures as inferior in terms of one's own norms and values is
    A. cultural relativity.
    B. scientific rationalism.
    C. ethnocentrism.
    D. linguistic relativity.

11. The ability to suspend judgment about other cultures in an attempt to get along in other parts of the world is known as
    A. ethnocentrism.
    B. cultural relativity.
    C. linguistic relativity.
    D. nationalism.

12. When people in societies that were colonized were forced to learn the language of a conquering nation, they experienced forced
    A. compromise.
    B. culture shock.
    C. acculturation.
    D. resistance.

13. When a group has become "Americanized" into the culture of the United States, sociologically it has become
    A. normalized.
    B. nullified.
    C. isolated.
    D. assimilated.

14. When a culturally distinct people within a larger culture fails to assimilate fully or has not yet become fully assimilated, we say that it is
    A. a society.
    B. a civilization.
    C. a subculture.
    D. an ideology.

15. Culture refers to
    A. everything in the man-made environment.
    B. the ability to create advanced technology.
    C. the artistic aspects of civilization.
    D. the complex webs of relationships that make up a society.

16. Which of the following was the most important contributing factor to the maintenance of the Cherokee subculture?
    A. The physical isolation from the dominant culture caused by a river basin
    B. Economic destitution that caused the people to come together for survival
    C. The development of high-tech industries in the Oklahoma area where the Cherokee live
    D. The development of a written language

17. Ethnocentrism is
    A. desirable because it keeps a society together and maintains social order.
    B. being culture-bound since it involves seeing the world through your own "glasses."
    C. detrimental because it forms the basis for all conflicts.
    D. functional in that it fosters the competitiveness required for a capitalistic system.

ANSWER KEY

The following provides the answers and references for the practice test questions. Objectives are referenced using the following abbreviations: T = text, R = Telecourse Guide Reading, and V = Video.

Answers	Lesson Goals	Objectives	References
1. A	1	T1	Kornblum, p. 60
2. D	1	T2	Kornblum, p. 62
3. A	1	T3	Kornblum, p. 62
4. D	1	T4	Kornblum, p. 62
5. C	1	T5	Kornblum, p. 66
6. D	1	T6	Kornblum, p. 66
7. C	2	T7	Kornblum, p. 69
8. C	2	T8	Kornblum, p. 69
9. B	2	T9	Kornblum, p. 71-72
10. C	3	T10	Kornblum, p. 72
11. B	3	T11	Kornblum, p. 73
12. C	3	T12	Kornblum, p. 77
13. D	3	T13	Kornblum, p. 78-80
14. C	3	T14	Kornblum, p. 80
15. A	4	V1	Video
16. D	4	V2	Video
17. B	4	V3	Video

Notes and Assignments:

Lesson 5

Socialization

LESSON ASSIGNMENT

Review the following assignment in order to schedule your time appropriately. Pay careful attention; the titles and numbers of the textbook chapter, the telecourse guide, and the video program may be different from one another.

Text:

Kornblum, *Sociology in a Changing World*,
Chapter 5, “Socialization,” pp. 124-155.

Reading:

There is no Reading for this lesson.

Video:

“Socialization,”
from the series, *The Sociological Imagination*.

OVERVIEW

When you were a child you interacted with those around you countless times. At first you declared who you were with coos, gurgles, cries, and occasional smiles, then with your first tentative words. Eventually, your verbal sentences were succeeded by written words. You began to scratch your impression on the universe. The planet would never be the same.

You were a combination of harmony and dissonance, of structure and change. You became an unrepeatable, unmatchable, and irreplaceable person. But you also became someone who is a participating member of your society. The fact that you are here now, reading these words, taking a sociology course, means that you know the language and expectations of others in your social world. You conform and contribute to your society. To paraphrase Clyde Kluckhoun, every person in some respects is like all others, like some others, and like no other.

When you think of yourself, it is likely to be as both conforming and conventional, as well as individual and creative. This total visualization, all the impressions and thoughts you have about you, is what sociologists call your *self.* But how did you get this self-image that dominates most of what you do each day?

In this lesson you discover how the biological organism that emerged writhing and gushing in the hands of the one who delivered you became transformed into a maturing adult by learning, roletaking, and identifying—through the process of social interaction.

This *socialization* process of becoming both a participating member of society and an individual begins in the microenvironments of home and neighborhood. But it quickly moves to the larger social environments of groups, organizations, and institutions.

Americans like to think of themselves as individuals. American culture encourages that. But the fact is that we are, to a great extent, reflections of our families, schools, peer groups, mass media, and other groups in our communities. This lesson examines the way each of these is instrumental in forming our personalities.

Right now, stop reading this overview and write at least ten answers to this question: "Who am I?"

Now inspect your list. Did you list man or woman? If not, it may be that this aspect of your self is so fundamental to your identity you take it for granted. You find out about gender identity as you study this lesson.

As you explore your list more carefully, try to identify two or more items in it that contradict or conflict with each other. Is there a role or quality that doesn't fit perfectly with the others on the list? If so, it may be because you have experienced discontinuities during the process of your socialization.

Socialization is one of the foundation concepts of sociology. If you study this lesson and complete the suggested assignments in the "Application Exercises" sections, you unearth a treasury of knowledge about your self.

LESSON GOALS

Upon completing this lesson, you should be able to:

1. Recognize how individuals become fully participating members of their society through the socialization process.
2. Analyze how the personality is socially constructed, applying sociological theories that emphasize interaction between the individual and social groups.
3. Relate how social environment is related to early socialization.
4. Analyze how the various agents of socialization affect the socialization process through the life course.
5. Analyze gender socialization, the social and cultural aspects of emotions, and emotional expression.

TEXTBOOK OBJECTIVES

The following textbook objectives are designed to help you get the most from the text. Review them, then read the assignment. You may want to write notes to reinforce what you have learned.

Text: Kornblum, *Sociology in a Changing World*, Chapter 5.

1. Define socialization and its three major phases.
2. Explain the nature vs. nurture controversy. Discuss the Freudian Revolution, behaviorism, studies of isolated children, and revelations concerning how the need for love impact this controversy.
3. Determine what Ivan Pavlov contributed to behaviorism.
4. Explain what sociologists mean when they speak of the "self" and "the social construction of the self."

5. Define the "looking glass self" (Cooley) and discuss what Mead revealed about the formation of self.

6. Identify a criticism of the findings reported in *The Bell Curve*.

7. Explain the importance of "taking the roles of others" in the socialization process from George Mead's theory. Describe "significant others" and "the generalized other." Recognize examples of each.

8. Define "face work." Explain how Erving Goffman use this concept.

9. Identify what Jean Piaget and Laurence Kohlberg contributed to the understanding of child development.

10. Recognize the trends and consequences of the socialization of homeless children on the streets.

11. Define agencies and agents of socialization. Determine what impact family, schools, community, peer groups, and mass media have on socialization.

12. Explain how socialization is impacted later in life by adult socialization and resocialization. Discuss the importance of "core identity".

13. Determine what Erikson contributed to our knowledge about lifelong socialization.

14. Differentiate among these terms: gender, sex, and gender identity.

15. Explain how ideas of manliness and womanliness affect people in meeting the demands of a changing world. Discuss Ron Kovic's illustration regarding gender socialization.

16. Identify what sociologists say about the effect of culture and social structure on emotions and emotional expression.

VIDEO OBJECTIVES

The following video objectives are designed to help you get the most from the video segment of this lesson. Review them, then watch the video. You may want to write notes to reinforce what you have learned.

Video: "Socialization"

1. Define the self and how it arises.
2. Determine the importance of the peer group in adolescence. Discuss why adolescence is considered one of the most crucial periods of socialization.
3. Describe the important agents of socialization in adulthood. Describe both socialization in the last chapters of peoples' lives and the "completion" of the self. Include how individuals are involved in their own socialization.

RELATED ACTIVITIES

These activities may be used by your instructor as written assignments or as discussion topics. They may also be included as essay questions on your tests.

1. Describe a situation in which your parents or guardians interacted with you and taught you a social skill that you have found to be helpful in your life. Why was it important to your life?
2. List and explain three social skills you learned from your parent(s); how did you learn each skill?
3. On the job—at work—you acquire knowledge, skills, and values, all of which help you to be a participating member of society. Briefly describe at least five things you have learned from those with whom you have worked that have increased your ability to participate in society.
4. The "looking glass self" has to do with the degree of self-esteem you have, based on your perceptions of the judgments of others about you. Some of these perceptions correspond to the actual evaluations of others; some do not.

Make a list of two or three of your significant others—either dead or alive—and, to the right of their names, describe what *you think* their assessments of you are. For each case, describe the reasons for your assessment.

5. Describe five things you learned from your schools that have helped you to survive in your society and to participate effectively in your community. List at least two things that could have helped you that you regret you did not learn at school. Explain why you feel this way.

6. Peer groups have both positive and negative influences on us. Explain how your peers have both helped and harmed you as a person.

7. Think of books, magazines, or other reading materials you were exposed to as a child. Analyze the content of those reading materials in terms of gender socialization. What pictures, photos, words, or examples perpetuated in you the idea of separate gender roles? Summarize your findings. If you can, describe ways that you took on the gender roles presented in the materials.

PRACTICE TEST

The following items will help you evaluate your understanding of this lesson. Use the answer key at the end of the lesson to check your answers or to locate material related to each question.

Multiple-Choice

Select the one choice that best answers the question.

1. Which of the following is true of socialization?
   A. It is based on genetic programming.
   B. It makes possible the transmission of culture from one generation to the next.
   C. It is the formal process whereby groups form and become social.
   D. It is a learning process that begins at about age five and ends around age sixty-five.

2. Which of the following best represents what most scientists today recognize about the “nature” or “nurture” controversy?
   A. The “nature” side of the debate is generally accepted as the correct position.
   B. Biological factors are dominant in the development of the personality.
   C. An individual’s development is an outcome of both innate characteristics and the many influences of his or her social environment.
   D. Biological factors in the development of personality are completely negligible due to the enormous influence of family during early childhood.

3. The findings of social and behavioral scientists studying the role of love in the development of the individual show that
   A. humans are inherently selfish and cruel.
   B. if found by late adolescence, children reared in extreme isolation can learn and develop in the same way as children reared in a healthy and nurturing environment.
   C. human infants cannot learn as fast as other primate infants in similar circumstances.
   D. nurturance and parental love play a profound, though still poorly understood, role.

4. The ways in which the self develops through social interaction in childhood and throughout life are captured in which of the following:
   A. personality development.
   B. self development.
   C. interactional formation.
   D. social construction.

5. As we mature, the overall pattern of our impressions of other people’s opinions becomes a dominant aspect of our own identities.

   This statement represents a concept of personality development referred to as
   A. “looking glass self.”
   B. behaviorism.
   C. Freudian.
   D. opinionism.

6. George H. Mead believed that the self is
   A. present at birth and is "brought out" by socialization.
   B. acquired by observing and assimilating the identities of others.
   C. evident prior to the development of language.
   D. dependent on the sexual bonding of the child with the opposite-sex parent.

7. According to Mead, people who loom large in our lives, after whom we model our behavior or whose behavior we seek to avoid, are called
   A. generalized others.
   B. peer groups.
   C. model groups.
   D. significant others.

8. According to the textbook, what tends to heighten the difficulties people have in the roles they have been socialized to perform?
   A. Cultural lag
   B. Social change
   C. "Face work"
   D. Cultural structures

9. According to Lawrence Kohlberg, which of the following is NOT a characteristic of moral development?
   A. Preconventional
   B. Middle-conventional
   C. Conventional
   D. Postconventional

10. According to your textbook, a major consequence of the high rate of teenage fertility in the United States is
    A. the continuation of school for the mother (and father) in "special," segregated classes.
    B. proportionately more babies are born into homes in which they will not receive the attention and material benefits that will help them realize full genetic potential.
    C. the "grandparents" become the child's "parents."
    D. the dramatic increase of teenage mothers enrolling in parenting classes to develop the skills necessary to be an effective parent.

11. Groups of people, along with the interactions that occur within those groups, that influence a person's social development are known as
    A. cultures.
    B. agencies of socialization.
    C. collectivities.
    D. shaping structures.

12. What is the primary agency of socialization?
    A. School
    B. Church
    C. Community
    D. Family

13. Interacting groups of people of about the same age are called
    A. age cohorts.
    B. peer groups.
    C. cohesive groups.
    D. secondary groups.

14. Which of the following is NOT a condition under which children's peer groups are more likely to form gangs?
    A. There are high levels of poverty.
    B. Handguns are readily available.
    C. Mass media is widely used.
    D. Adolescents seek protection and companionship in what is perceived as a hostile world.

15. The part of the self formed in early childhood that does not change easily is called
    A. core identity.
    B. personal self.
    C. personality.
    D. extroverted self.

16. According to Erik Erikson, at which stage does one experience the conflict between intimacy and isolation in the quest for love?
    A. Adulthood
    B. School age
    C. Play age
    D. Early adulthood

17. The ways in which we learn gender identity and develop according to cultural norms of "masculinity" and "femininity" are known as
    A. gender socialization.
    B. sexual identification.
    C. identity change process.
    D. masculine/feminine development.

18. Why so many girls become women who are unhappy with their bodies and boys grow into men who are happy with their bodies is
    A. found in an analysis of American culture.
    B. perceived as a mystery to social science.
    C. based on biological traits over which boys and girls have no power.
    D. based on the genetic aggressiveness of boys, who are therefore in a more dominant position with regard to personality development.

19. Most sociologists would argue that differences in emotional behavior between the sexes are yet another example of the results of
    A. genetic programming.
    B. innate ability of women to nurture.
    C. social-class distinctions.
    D. gender socialization.

20. The self is the individual's ability to
    A. read and write using complex words and phrases.
    B. get outside one's own skin and to have some sense of how others are responding to him or her.
    C. become independent and to make it in the world without help from parents or other relatives or from guardians.
    D. recognize feelings that are one's own.

21. During adolescence the peer group becomes
    A. less important than church as an agency of socialization.
    B. a shelter from family abuse and usually substitutes for the family.
    C. important in that teenagers learn sexuality and what makes them attractive to others.
    D. less significant in the socialization process than are the mass media.

22. In adulthood, the primary agents of socialization are more focused on
    A. media images and role models.
    B. colleagues at work, husband or wife, and close friends.
    C. immediate family than in previous years.
    D. religious affiliations.

ANSWER KEY

The following provides the answers and references for the practice test questions. Objectives are referenced using the following abbreviations: T = text, R = Telecourse Guide Reading, and V = Video.

Answers	Lesson Goals	Objectives	References
1. B	1	T1	Kornblum, p. 126
2. C	1	T2, T3	Kornblum, pp. 131-133
3. D	1	T2, T3	Kornblum, p. 131
4. D	2	T4	Kornblum, p. 133
5. A	2	T5, T6	Kornblum, p. 133
6. B	2	T5, T6	Kornblum, p. 133
7. D	2	T7	Kornblum, p. 134
8. B	2	T8	Kornblum, p. 135
9. B	2	T9	Kornblum, p. 136
10. B	3	T10	Kornblum, p. 138
11. B	4	T11	Kornblum, p. 139
12. D	4	T11	Kornblum, p. 139
13. B	4	T11	Kornblum, pp. 141-142
14. C	4	T11	Kornblum, p. 143
15. A	4	T12	Kornblum, p. 144
16. D	4	T13	Kornblum, p. 146, Table 5.1
17. A	5	T14	Kornblum, p. 146
18. A	5	T15	Kornblum, p. 147
19. D	5	T16	Kornblum, p. 146
20. B	5	V1	Video
21. C	5	V2	Video
22. B	5	V3	Video

Notes and Assignments:

Lesson 6

Groups

LESSON ASSIGNMENT

Review the following assignment in order to schedule your time appropriately. Pay careful attention; the titles and numbers of the textbook chapter, the telecourse guide, and the video program may be different from one another.

Text:

Kornblum, *Sociology in a Changing World*,
Chapter 6, "Interaction in Groups," pp. 156-174.

Reading:

Telecourse Guide for The Sociological Imagination, pp. 76-78.

Video:

"Groups,"
from the series, *The Sociological Imagination.*

OVERVIEW

How many times have you picked up the paper to read about a mass killer, an assassination attempt, or a murder-suicide, only to discover that the perpetrator was a loner, a person isolated from neighbors and community, with few if any friends; and someone alienated from family members? These sad and unfortunate cases point up how important social groups are for all of us.

Certainly, some people are more gregarious than others. But social interaction within some sort of group or groups is absolutely essential for social and psychological survival, as well as for healthy functioning within the social system. Americans are known as joiners; however, all societies depend on social groups as the basic elements, or building blocks, of their social systems. These groups take many forms: cliques, couples, corporations, clubs, and kin.

Sociologists have always been interested in social groups, and that is what this lesson is about. You find out there are different kinds of groups and what makes them tick. You may even resolve your questions about situations in which you have been included or excluded from a social group. What is involved in this process? Why do groups place boundaries around themselves? What are the different forms these boundaries take?

We interact with perhaps hundreds of people each day. Strangers pass in the hallway, nodding their heads to one another. Two drivers angrily exchange honks. Lovers kiss. A crowd hisses at a poor performance by a music group. All of these are interactions, but not all of them represent social relationships. As we interact, we sometimes form relationships. Here we examine the different kinds of social bonds we form.

Have you ever become irritated with friends because they weren't listening to you, were taking advantage of your generosity, or perhaps were not giving you what you needed in the relationship? As we interact, we follow certain unwritten guidelines or principles. This lesson lets us scrutinize that interaction process and try to understand why we do certain things in certain situations. Finally, we look at the different kinds of leadership that emerge within groups.

Doesn't it sound fascinating? Social interaction, relationships, and social groups. They are a focus of much sociological research and thinking. Perhaps you will even start seeing yourself and your own social situations in a different light once you are captivated by these sociological insights.

LESSON GOALS

Upon completing this lesson, you should be able to:

1. Analyze social groups: what different kinds exist, how they function, how they differ from other social entities that are not groups, and how they may exclude or include people as members.

2. Analyze social relationships within groups, the ways that people form and use social bonds, and analyze how groups distinguish or set themselves off from one another.

3. Analyze interaction within and between groups, how this interaction contributes to social stability, and how it brings about changes within groups and society.

4. Analyze the significance of groups in society and how our lives are influenced by our membership in the groups. Analyze the interplay between conformity and individualism.

TEXT OBJECTIVES

The following textbook objectives are designed to help you get the most from the text. Review them, then read the assignment. You may want to write notes to reinforce what you have learned.

Text: Kornblum, *Sociology in a Changing World*, Chapter 6.

1. Define a social group.
2. Determine the characteristics of social groups.
3. Describe the relationships in and the characteristics of a secondary group.
4. Explain and exemplify how social networks can exist in and well beyond a neighborhood.
5. Explain the meaning of in-groups and out-groups. Be able to recognize examples.
6. Define a reference group. Recognize examples of reference groups.
7. Explain and exemplify the pleasure principle.
8. Describe the rationality principle. Recognize examples of the rationality principles.
9. Describe and exemplify the reciprocity principle.

10. Define the fairness principle. Recognize examples of fairness principle. Explain how it may outweigh the rationality principle in certain situations.

11. Describe impression management. Explain the difference between "frontstage" and "backstage" behavior. Recognize examples of each.

12. Describe the bystander effect and its sociological causes.

13. Differentiate between the instrumental/task leader and the expressive/socioemotional leader. Be able to recognize examples of each.

READING
WHAT A SOCIAL GROUP IS AND WHAT IT ISN'T

READING OBJECTIVES

The following reading objectives are designed to help you get the most from the Reading. Review them, then read the assignment. You may want to write notes to reinforce what you have learned.

Reading: *Telecourse Guide for The Sociological Imagination*, pp. 76-78.

1. Define a social category, aggregate, and collectivity, and differentiate between these and a social group. Be able to recognize examples of each of the above.

2. Describe the characteristics and relationships within a primary group. Be able to recognize examples of primary groups.

The Social Groups in Our Lives

What do you think of when you hear the word "society"? Do you visualize your teachers or the police, see a map of your country, or think of "the government," "the establishment," or something even vaguer? The fact is, we experience society in the social groups to which we belong, that is, in our families, friendships, church groups, and clubs, as well as where we work.

These social groups are the basic units that form a social structure and keep society together. A *society* is a large, complex system of relationships existing within boundaries and satisfying the basic needs of members. Without groups there would be no society, just as there would be no human body without the billions of cell structures that constitute it. Social groups are where we live our lives; we could not exist without them.

Although we frequently speak of a social group in conversation, sociologists use the term in a special and specific manner. It is a basic concept in sociology, yet it is sometimes misunderstood. *Social groups consist of people who have coordinated social relationships that comprise a system of interdependence.*

You have probably heard it said that humans are social animals, however, not all social situations involve groups. At this moment you are acting as an undergraduate student. That is a social characteristic you share with students all over the world. Therefore, you are in the *social category* of college student. Social categories are not social groups.

During the day you may find yourself in a traffic jam or waiting in line for a donut or in a packed elevator. Although these are social situations where you are aware of others and may interact with them, here you are part of an *aggregate*, people who simply happen to be in the same place at the same time. Social categories and aggregates are *not* social groups because they are not organized in any way.

Other social entities that sociologists study are *collectivities.* These have a small degree of social structure and organization, so they lie between aggregates and social groups. Audiences and crowds exhibit collective behavior, so they are called *collectivities. Social movements*, such as the women's movement, also are collectivities. These movements are unusual collectivities in that they contain many social groups, but they also embrace individuals who support and sympathize with the social movement yet may not be a member of an organized group.

Unlike social categories, aggregates, and collectivities, *social groups* are systems of *organized relationships*. Persons in a family, for example, play roles: parent, child, dishwasher, cook, student, breadwinner, etc. They coordinate their actions to make a more or less cohesive whole. And people in a bureaucracy organize their actions through roles such as chief executive officer, manager, sales representatives, or secretary. A bureaucracy, therefore, is a social group.

A *social relationship* consists of a persisting pattern of interaction between two or more persons; one or two conversations with someone do not make a relationship. Also, you must maintain interaction or communication in order to continue a social relationship. Divorce or other relational breakups, for example, are often caused by poor or infrequent communication.

In small, simple societies, such as the bushpeople of the Kalahari, all social relationships are what we call *primary relationships* because they are marked by close bonds of attachment and personal, face-to-face interaction. Any group with this kind of relationship is a primary group. Since you live in a modern society, you not only have primary relationships, you also have countless *secondary relationships*. These are the more formal, impersonal relationships in which you and others play specialized, but more limited, roles. With these, much of the interaction may even be based on formalized contracts. You have a secondary relationship with the president of the United States, and you also may have one with an oil company symbolized by your plastic credit card with an account number.

Groups have *norms*, that is, rules or expectations that govern and guide their behavior. Americans should drive on the right-hand side of the road. It is these "shoulds" and "should nots," the specific behavioral prescriptions and proscriptions that people carry in their minds, that make possible social life as a whole, as well as orderly interaction within each group. All groups, therefore, create some of their own norms.

The main thing to remember about social groups is that they are people who are organized! Don't mistake the term "social organization" to refer only to formally structured groups. By definition, *all* social groups, even a children's play group, have some degree of social organization.

VIDEO OBJECTIVES

The following video objectives are designed to help you get the most from the video segment of this lesson. Review them, then watch the video. You may want to write notes to reinforce what you have learned.

Video: "Groups"

1. Explain the significance of groups to individuals and to society. Be able to recognize examples.

2. Recognize examples of how groups give meaning to life, yet require commitment and sacrifice. Identify ways groups exert their influence on individuals and bring about greater commitment by requiring members to renounce their own ideas or lifestyles in favor of those advocated by the group. Determine different methods various groups use to try to erase individual differences in order to promote group unity.

3. Identify the factors that determine the degree of an individual's commitment to the group. Determine the risks involved in group pressures on individuals to conform.

4. Identify the (leadership) qualities necessary for someone to be an effective group leader today.

RELATED ACTIVITIES

These activities may be used by your instructor as written assignments or as discussion topics. They may also be included as essay questions on your tests.

1. Name two primary groups to which you belong. Primary groups can sometimes be dysfunctional (harmful) to individuals. Cite one or two instances you know of in which this was the case.

2. Name a secondary group to which you belong or a secondary relationship that you have.

3. Explain and give examples of primary and secondary groups.

4. Describe an example of boundary maintenance of a group you or someone you know belongs to.

5. Name an in-group to which you belong, and describe a situation in which you were excluded from an out-group.

6. Name a reference group to which you belong and have membership, and name a reference group to which you do *not* presently belong.

7. Describe a situation in which you decided *not* to join a group or interact with someone as a friend because the "losses" would be too great.

8. Over a span of several days, look for "frontstage" and "backstage" behavior in your family, at school, or at work. You may have to become an eavesdropper; however, do not violate anyone's privacy.

PRACTICE TEST

The following items will help you evaluate your understanding of this lesson. Use the answer key at the end of the lesson to check your answers or to locate material related to each question.

Multiple-Choice

Select the one choice that best answers the question.

1. A realization about who belongs and who does not belong implies that group members have a sense of the
   A. size of their group.
   B. boundaries of their group.
   C. density of their social relationships.
   D. number of dyads in their group.

2. A social group is a set of people who
   A. merely happen to be in the same place at the same time.
   B. are involved in collective behavior.
   C. are bound together by a set of membership rights and mutual obligations.
   D. simply have some social characteristic in common.

3. Groups whose members have strong positive attachments to each other are said to be highly
   A. specialized.
   B. alienated.
   C. cohesive.
   D. political.

4. Secondary groups are usually based on
   A. small size.
   B. emotional involvements.
   C. intimacy.
   D. some form of contract.

5. The relationships the textbook describes between the Nortons and local political party organizers, as well as with racketeers, demonstrates
   A. reference group association.
   B. social networking outside territorial communities.
   C. primary relationships becoming a collectivity.
   D. non-identification with the secondary group, the Nortons.

6. One's own peers and close associates are one's
   A. secondary group.
   B. out-group.
   C. in-group.
   D. collectivity.

7. Gilda wants to be a lawyer. She doesn't know any lawyers personally, but she knows a lot about the profession and often thinks about herself in the role of lawyer. She has decided to take courses in undergraduate school that will help her when she goes to law school.

   For Gilda, lawyers are a
   A. primary group.
   B. reference group.
   C. social network.
   D. cohesive group.

8. Alex kept insisting that the group play the game by the book and continually questioned people's moves. What had started out as fun became tedious and tense. Gradually people began to make excuses and depart. In reality, they were just not enjoying the game any more.

   Alex's associates were acting on the principle of
   A. reciprocity.
   B. pleasure.
   C. fairness.
   D. rationality.

9. When people make rough calculations of costs and benefits, as well as profit and loss, in their interactions with others, they are acting out of the
   A. reciprocity principle.
   B. rationality principle.
   C. fairness principle.
   D. pleasure principle.

10. What others do for you, you should try to do for others is a statement of the
    A. reciprocity principle.
    B. pleasure principle.
    C. fairness principle.
    D. rationality principle.

11. Which principle of interaction is being violated when rules are not being applied equally or when one is not rewarded in the same ways as others?
    A. Reciprocity
    B. Pleasure
    C. Fairness
    D. Rationality

12. The movie star presented herself as congenial and kind in the interview on the talk show. After the interview, she showed her true colors by scolding the cameraman for getting too close and yelling at the producer about the amount of time she had on camera.

    The star's behavior when the cameras were off is an example of
    A. impression management.
    B. secondary group attachment.
    C. frontstage behavior.
    D. backstage behavior.

13. When people try to hide behind their anonymity and not get involved, they are exhibiting
    A. impression management.
    B. bystander effect.
    C. altruism.
    D. primary group attachment.

14. A group leader who is concerned with strict adherence to group norms and who takes the lead in carrying out group goals is
    A. a socioemotional leader.
    B. a task leader.
    C. an authoritarian leader.
    D. a strategic leader.

15. People waiting on a street corner at the "don't walk" light are
    A. a social group.
    B. an aggregate.
    C. a social category.
    D. a primary group.

16. Most primary groups are characterized by
    A. face-to-face relationships with close bonds.
    B. specialized roles.
    C. impersonality.
    D. large size.

17. Janet and Gus have known each other for two years. They meet every Wednesday night for long discussions and have developed a close relationship.

    Gus and Janet represent a
    A. primary group.
    B. secondary group.
    C. social category.
    D. population.

18. Most things in the modern world get done by
    A. individuals acting alone.
    B. social categories.
    C. groups.
    D. collectivities.

19. Which of the following does NOT illustrate the commitment and sacrifice sometimes required by groups?
    A. "Having had children, I do not have much private time."
    B. "The children are five and seven years old and have blue eyes."
    C. "My husband and I don't have much time for just the two of us since we had children."
    D. "We don't have some of the material possessions like nice furniture and a nice car."

20. The degree of individual commitment to a group is related to
    A. knowledge about the process of group formation.
    B. importance to the individuals, their livelihood, and their safety.
    C. diversity in the gender composition of the group.
    D. number of people, that is, an odd or even number.

21. According to the video program about groups, one of the most underdeveloped leadership skills is that of
    A. physical agility.
    B. thinking.
    C. listening.
    D. efficiency.

ANSWER KEY

The following provides the answers and references for the practice test questions. Objectives are referenced using the following abbreviations: T = text, R = Telecourse Guide Reading, and V = Video.

Answers	Lesson Goals	Objectives	References
1. B	1	T1	Kornblum, p. 158
2. C	1	T1	Kornblum, p. 158
3. C	1	T2	Kornblum, p. 159
4. D	1	T3	Kornblum, pp. 159-160
5. B	2	T4	Kornblum, p. 161
6. C	2	T5	Kornblum, p. 161
7. B	2	T6	Kornblum, p. 164
8. B	3	T7	Kornblum, p. 168
9. B	3	T8	Kornblum, p. 168
10. A	3	T9	Kornblum, p. 168
11. C	3	T10	Kornblum, p. 169
12. D	3	T11	Kornblum, pp. 171-172
13. B	3	T12	Kornblum, p. 172
14. B	3	T13	Kornblum, p. 174
15. B	1	R1	TG Reading
16. A	1	R2	TG Reading
17. A	1	R2	TG Reading
18. C	4	V1	Video
19. B	4	V2	Video
20. B	4	V3	Video
21. C	4	V4	Video

Notes and Assignments:

Lesson 7

Formal Organizations

LESSON ASSIGNMENT

Review the following assignment in order to schedule your time appropriately. Pay careful attention; the titles and numbers of the textbook chapter, the telecourse guide, and the video program may be different from one another.

Text:

Kornblum, *Sociology in a Changing World*,
Chapter 6, "Interaction in Groups," pp. 174-189.

Reading:

Telecourse Guide for The Sociological Imagination, pp. 90-92.

Video:

"Formal Organizations,"
from the series, *The Sociological Imagination.*

OVERVIEW

Donna's frustration and anger were palpable. Her face was red and the veins in her neck seemed to bulge and throb. She steamed inside over the coldness and robotlike manner of the young man in the registrar's office who asserted, "Those are the rules. You can't register until I see the receipt for the fine on that overdue book from last semester."

Donna knew she had mailed the check. She persisted in trying to drive home her case, showing the bureaucrat the entry in her checkbook. "See, I made the check out two months ago," she said.

"Yeah, but you made it out to financial aid and not to the library, lady. That's the problem. It's not my fault. You don't need to get mad at me! I'm just trying to do my job."

Then Professor Hatfield happened to walk by. He recognized Donna, one of his brightest students, and could see that she was having some kind of difficulty. When he found out what it was, he called Juanita Hall, the director of financial aid, who immediately recognized his voice and said she'd be glad to search for the check or some record of it.

In five minutes Mrs. Hall was back on the line, apologizing for the fact that the check had been misplaced and saying that she now had found it. She thanked the professor for the call and assured the registrar clerk that he could go ahead and process Donna.

Within an hour Donna had completed her registration and was ready to continue to pursue her cherished goal of becoming an engineer.

Donna had just encountered a bureaucracy. She experienced its impersonality, its rules and set procedures that apply the same to everyone, its dysfunction and its efficiency, and the informal structure that helps it to complete its assigned tasks.

One of the most important kinds of social structure in modern societies is the formal organization. Most of us could not survive without these organizations. They mark the urban landscape like the skyscrapers that often house them. They permeate social life from the city and the suburb to the small rural settlement whose central institution is the general store that also houses the post office, where many of the older citizens of the area come to pick up their monthly Social Security checks.

We all benefit from one of the main kinds of formal organization, bureaucracies. When we go to church or attend a meeting of a charitable association we belong to, when we work for a company or write a letter to the newspaper, or even when we hang our clothes in the locker at the health club, we are involved with formal organizations. Much of the economic activity that takes place in the world happens within the confines of formal organizations.

This lesson unfolds for your sociological imagination the story of these sometimes beguiling, sometimes disconcerting, social systems. In the video program you see two American companies that are trying to reformulate their structures and procedures to better reach their goals, serve their customers, and encourage their employees.

Some of your assumptions about this subject may be challenged in this lesson. But that is one of the goals of sociology—to transform consciousness.

LESSON GOALS

Upon completing this lesson, you should be able to:

1. Analyze formal and informal organizations, and describe the meaning and importance of voluntary organizations.

2. Analyze bureaucracy, including Max Weber's description of its typical characteristics, and the relationship between bureaucracy and rationality.

3. Analyze the relationship between the individual and bureaucracy.

4. Analyze the dysfunctional aspects of bureaucracy and the alternatives being formulated by various organizations.

5. Analyze examples of formal organizations to show how they are bureaucracies, yet how they are changing.

TEXTBOOK OBJECTIVES

The following textbook objectives are designed to help you get the most from the text. Review them, then read the assignment. You may want to write notes to reinforce what you have learned.

Text: Kornblum, *Sociology in a Changing World*, Chapter 6.

1. Explain the meaning of formal organizations, the forms they take, and how they are different from informal organizations.

2. Define and give examples of voluntary associations.

3. Define bureaucracy, and describe how is it different from a voluntary association.

4. List and explain the typical aspects of bureaucracy as set out by Max Weber. Recognize examples of each.

5. Explain Max Weber's idea about the rationality of bureaucracy.

6. Describe the purpose and results of the experiments performed by Stanley Milgram. Determine what he discovered about the conditions under which conflict occurs.

7. Explain the findings of Morris Janowitz and Edward Shils about the source of commitment in the Nazi army. Discuss how the U.S. Army has implemented what it learned from this study.

READING
BEYOND BUREAUCRACY

READING OBJECTIVES

The following reading objectives are designed to help you get the most from the Reading. Review them, then read the assignment. You may want to write notes to reinforce what you have learned.

Reading: *Telecourse Guide for The Sociological Imagination*, pp. 90-92.

1. Describe the dysfunctional aspects of bureaucracies.

2. Discuss how organizations changing to avoid the dysfunctional aspects of bureaucratic structure. Give examples of the changes and of alternative organizations that have been forming.

Building a New Corporate Culture

No human group works perfectly. We have alluded to the strains or dysfunctions in bureaucracies in the Overview and in the textbook. The video program explores more of the difficulties. Even the aspects of bureaucracy that are supposed to help it be more efficient sometimes impede it.

For example, the specialization that occurs as a result of the clearly defined responsibilities may be accompanied by rigid thinking, an inability to change or adapt to unique situations and problems that call for creative solutions.

Although the impersonality of the bureaucracy helps it to be impartial and objective, those who work for a bureaucracy may forget that people come to it as whole persons with dignity; the so-called bureaucrats may treat a client or customer more as a number or a problem than as a person. Also, rules sometimes become an end in themselves, rather than a means to reach the goals of the organization. These are some of the dysfunctional aspects of bureaucracy.

Institutions and communities have attempted to avoid some of the problems and dysfunctions of bureaucracies by transforming their organizations or by building alternative organizations. Formal organizations created to meet the needs of members of the community, without the rules and hierarchy of the bureaucracy, are springing up all over. Cooperatives of all kinds and so-called free, or alternative, schools are examples.

Many companies and other groups have tried to restructure their organizations to maximize participation, creativity, innovation, productivity, and efficiency. This often is referred to as building a new corporate culture.

For example, companies have come to realize that attention to the welfare of workers both on and off the job pays healthy dividends. Here are some of the specific things these companies have tried to do:

1. Emphasize more participation by all employees in decision making, causing members of the organization to take more responsibility for the product or service they are providing because they develop a greater sense of investment in the organization. This more democratic approach makes rules and formal hierarchical divisions less important than in the bureaucracy.

2. Make individual employees more important than attempts to pressure and control them, and recruit people on the basis of their support of the goals of the organization. If members of the organization are not doing well or are unhappy with a job, they often are retrained, moved, or rotated to another position.

3. Think of the organization as a community of equals with a minimum of status distinctions, rather than as a formal organization. One company we know of doesn't alphabetize its company directory by titles, departments, or even last names; its listings are in alphabetical order by first names.

4. Recognize the importance of informal social relationships and of emotional and family needs. Provide an atmosphere in which these values flourish, one that produces incentives to work that mere material rewards often fail to do.

Formal organizations are changing—sometimes to survive, sometimes as the key to becoming wildly successful! But the most effective organizations embody many of the features described above. Much of this change has been motivated by competition from the Japanese or by losing good workers who have become dissatisfied in a culture where the job no longer provides the only or ultimate satisfaction. The findings of social scientists who have studied alternatives to the bureaucratic form are proving invaluable in constructing internal guideposts.

VIDEO OBJECTIVES

The following video objectives are designed to help you get the most from the video segment of this lesson. Review them, then watch the video. You may want to write notes to reinforce what you have learned.

Video: "Formal Organizations"

1. Explain how our lives are influenced by formal organizations and cite examples of this from the video program and in your own life.

2. Describe the significance of bureaucracies in various formal organizations. Describe the elements of bureaucratic organization that are found in the context of the university. Be able to recognize examples.

3. Describe the dysfunctions of bureaucracy. Give examples from the university.

4. Using IBM as an example, discuss how some large corporations are attempting to address the needs of their employees. Be able to recognize examples.

5. In the video program on formal organizations, there is a woman who left a Fortune 500 company to start her own business. Describe the negative aspects of the bureaucracy that she left. Identify what organizations can do to implement authentic respect for, caring about, and valuing of employees.

RELATED ACTIVITIES

These activities may be used by your instructor as written assignments or as discussion topics. They may also be included as essay questions on your tests.

1. List and explain three positive and three negative aspects of working within a large bureaucracy.

2. Describe a situation or incident in which you have experienced two or more of the typical aspects of bureaucracy. Explain how these traits were beneficial and helped you achieve your goals.

3. Describe the findings of Morris Janowitz and Edward Shils about the importance of primary groups in armies.

4. Describe a situation or incident in which you experienced the dysfunctional aspects of a bureaucracy.

5. List at least one formal organization that has affected you during each of these periods: early childhood, elementary school, high school, and college years.

6. Based on the comments made by the employees of IBM Corporation and of Boeing Aerospace and Electronics, identify which of the two companies attracts you more. Explain why.

PRACTICE TEST

The following items will help you evaluate your understanding of this lesson. Use the answer key at the end of the lesson to check your answers or to locate material related to each question.

Multiple-Choice

Select the one choice that best answers the question.

1. Formal organizations have
   A. few if any written rules.
   B. relationships that arise spontaneously out of interaction among primary groups.
   C. leadership that comes from the primary group relationships and is based on the norms of those relationships.
   D. explicit, often written, roles that specify each member's relationships to others in the group.

2. A group that people join to pursue interests they share with other members of the group is known in sociology as
   A. a bureaucracy.
   B. a dominant group.
   C. a voluntary association.
   D. an authoritarian group.

3. In bureaucracies power is generally based on
   A. majority rule.
   B. moral principles.
   C. democratic principles.
   D. executive orders.

4. According to Max Weber, in a bureaucracy, the ideal official conducts business with a spirit of
   A. formalistic impersonality.
   B. passion.
   C. affection.
   D. enthusiasm.

5. According to Max Weber, which of the of the following is not a way bureaucracies "rationalize" human societies?
   A. Democracy
   B. Impersonality
   C. Norm of efficiency
   D. Rules

6. The experiments of Stanley Milgram were
   A. designed to take a close look at the act of obeying.
   B. observe people in their informal groups.
   C. test the rule of efficiency in formal organizations.
   D. discover the level of emotions under which people operate with greatest productivity.

7. In their study of German soldiers, Morris Janowitz and Edward Shils found that soldiers would continue to fight even in the face of defeat because of their
   A. fear of the wrath of their superiors.
   B. commitment to Nazi ideology and their belief in its superiority.
   C. excellent training and superior self-discipline.
   D. loyalty to their small combat units and their devotion to the other members of the units.

8. Which of the following is a dysfunctional aspect of bureaucracy?
   A. Use of majority rule in a formal structure
   B. Informal organizations within the bureaucracy
   C. Forgetting that people being served have dignity
   D. Adherence to the norm of efficiency

9. Which of the following is a way some formal organizations are restructuring as an alternative to bureaucracy?
   A. Attention to the welfare of workers
   B. Use of rules and precedents
   C. Positions ordered in a hierarchy
   D. Use of more authoritarian systems of decision making

10. Which of the following does NOT describe a way our lives are influenced by formal organizations?
    A. Formal organizations usually have lifespans longer than the people who comprise them.
    B. Formal organizations control our primary group relationships without our realizing they are doing so.
    C. Formal organizations are introduced so gradually that we easily adjust to them.
    D. We are almost unaware of how we are bound by the rules and regulations of formal organizations.

11. Which of the following is an example of the elements of bureaucracy in the context of the university?
    A. Labor is divided among people who deal with maintenance, provide instruction, and handle enrollment.
    B. People must staff departments that apply for grants from various public and private agencies and foundations.
    C. Nursery staff must perform their function of caring for the children of the staff and students.
    D. Committees that evaluate the match of values between community and procedures of the university are appropriately staffed.

12. Students concentrating on passing exams, obtaining good grades, and satisfying degree requirements but never really getting an education provide an example of
    A. dominance.
    B. impersonality.
    C. dysfunctional aspect of bureaucracy.
    D. division of labor.

13. Large corporations such as IBM have used various strategies to try to make their organizations work better while basically retaining a bureaucracy.

    Which of the following is NOT mentioned in the video program as one of these strategies or methods?

    A. Organizing decision making in the form of a reverse pyramid, that is, ideas come from the lowest employees to the highest.
    B. Attempting to involve different levels of the organization in decision making.
    C. Making employees more important and more a part of the corporate family.
    D. Having "respect for the individual" as a basic belief and value.

14. Which of the following do large corporations such as IBM typically use to achieve greater involvement by their employees?

    A. Greater adherence to Max Weber's ideal type of bureaucracy
    B. Involvement of employees in leveraged buy-outs
    C. Encouraging employees to communicate their opinions through surveys, such as a "speak up" program, and by an open door policy
    D. Encouraging employees to bring their children to work to see what their parents do when away from home

15. A large-scale organization desiring greater efficiency and effectiveness will emphasize authentic respect and caring for, and valuation of, employees.

    Which of the following best illustrates that respect, caring, and valuation?

    A. The organization establishes a strong hierarchy of authority in which the chain of command is made clear to employees so there is no confusion.
    B. Greater use of computers makes it possible to communicate tasks more efficiently and to check on the productivity of workers.
    C. Individual values are subsumed by corporate values.
    D. The organization becomes enthusiastic about democratization of the work culture.

ANSWER KEY

The following provides the answers and references for the practice test questions. Objectives are referenced using the following abbreviations: T = text, R = Telecourse Guide Reading, and V = Video.

Answers	Lesson Goals	Objectives	References
1. D	1	T1	Kornblum, p. 174
2. C	1	T2	Kornblum, pp. 174-175
3. D	2	T3	Kornblum, p. 175
4. A	2	T4	Kornblum, p. 175
5. A	2	T5	Kornblum, p. 175
6. A	3	T6	Kornblum, p. 177
7. D	3	T7	Kornblum, p. 178
8. C	4	R1	TG Reading
9. A	4	R2	TG Reading
10. B	5	V1	Video
11. A	5	V2	Video
12. C	5	V3	Video
13. A	5	V4	Video
14. C	5	V4	Video
15. D	5	V5	Video

Notes and Assignments:

Lesson 8

Societies

LESSON ASSIGNMENT

Review the following assignment in order to schedule your time appropriately. Pay careful attention; the titles and numbers of the textbook chapter, the telecourse guide, and the video program may be different from one another.

Text:

Kornblum, *Sociology in a Changing World*,
Chapter 4, "Societies and Nations," pp. 90-121.

Reading:

There is no Reading for this lesson.

Video:

"Societies,"
from the series, *The Sociological Imagination*.

OVERVIEW

Chaco pressed his point: "We are a society and a state. We have our own land—our reservation and our Navajo culture are in so many ways just the opposite of yours. Our language, our art, our customs, and our values do not fit into the good ol' American materialistic, earth-manipulating mentality."

"But, Chaco, you are my friend," Henry responded. "You speak English; we are in a lot of the same groups at college; you're one of us."

"And besides," Henry continued, "don't your people have to obey American laws? Can't they vote in American elections? You're all American citizens. Doesn't that mean you are Americans—part of the American society—rather than a separate Navajo society?"

"Well, in a way," Chaco answered. "But on the reservation we have our own governmental bodies, our own laws and law-enforcement agencies. Doesn't that make us a "state" the way our sociology teacher explained it?"

Chaco and his Anglo friend Henry were confused. Where does the tribe stop and the society begin? How does one distinguish society from the state or a nation-state?

Chaco was certainly part of such diverse social structures as friendship cliques, his five college classes, and the college itself. Yet he never really felt that he fit in. He was secretly uneasy when his friends asked him why he didn't have the latest electronic gadgets or a car.

Unlike his fellow Navajos on the reservation, Chaco could afford these things; he simply didn't want them. And, in the middle of lectures, he often daydreamed and yearned to be outside under the trees, feeling the breeze cut through his long black hair. He missed his family and the expanse of the Arizona desert. In this reverie he was Navajo; it felt good and natural.

Yet Chaco liked Henry and his other friends, and he valued the learning opportunities at the college. It was just that his Indian nature was at war with his role as an American college student. The other students had difficulty with the duality, too. Even though Chaco was making an A in every course, his classmates sometimes patronized him by asking if he needed help with the language or his class work.

Chaco's Native American ancestors had lived a simple life. Some were hunters, some farmed, some were sheepherders. But the incursion of the burgeoning white civilization had transformed many of their lives. New technologies were used to overrun them, threatening their social, cultural, and physical existence.

The realities Chaco and his people faced are a microcosm of several ideas examined in this lesson. One of them involves the impact on an individual of the various levels of social structure, especially society, state, and nation-state. Another is illustrated by the several statuses and roles Chaco played and by the internal and external conflicts and changes which he experienced. We even see how the effects of population growth and technology on the history of Chaco's people relate to aspects of our own lives.

In the course of this lesson, you discover how macro-changes in society and throughout history affect you as an individual. You are challenged to examine how you fit into the different levels of social structure. And, like Chaco, you may experience conflicts arising from your various roles and social statuses. Perhaps this lesson can help you to understand and cope with them just a little better.

LESSON GOALS

Upon completing this lesson, you should be able to:

1. Analyze different elements and levels of social structures—especially societies—and how they adapt and change.

2. Analyze different kinds of societies and how they change over time in size, technology, and social structure.

3. Analyze how social relationships change as societies grow larger and how the individual is affected by social structure.

4. Analyze the difference between a nation and a society, and evaluate international social changes using this analysis.

5. Analyze how societies of varying complexity satisfy basic human needs at different places and times.

TEXTBOOK OBJECTIVES

The following textbook objectives are designed to help you get the most from the text. Review them, then read the assignment. You may want to write notes to reinforce what you have learned.

Text: Kornblum, *Sociology in a Changing World*, Chapter 4.

1. Explain the meaning of a society and how it differs from a population. Distinguish between societies and other social structures, and be able to recognize examples of societies.

2. Define statuses, and describe and exemplify how they change to adapt to new conditions.

3. Recognize examples of roles and role expectations.

4. Describe hunting and gathering societies, and explain how they were different from modern societies.

5. Describe agrarian societies and the new social structures that accompany them.

6. Discuss the changes in population patterns, technology, and societal structure that appeared with the industrial revolution.

7. Explain the difference between *gemeinschaft* and *gesellschaft* societies.

8. Describe how the role conflict and role strain relate to personal life changes.

9. Differentiate between ascribed and achieved status and be able to recognize examples of both.

10. Explain and exemplify how master status can be used as a means of discrimination.

11. Differentiate between the state and the nation-state.

12. Describe Max Weber's concept of the state and legitimacy.

13. Recognize examples in which there sometimes is not a clear match between society and nation. Describe problems that can arise in this situation.

VIDEO OBJECTIVES

The following video objectives are designed to help you get the most from the video segment of this lesson. Review them, then watch the video. You may want to write notes to reinforce what you have learned.

Video: "Societies"

1. Explain and exemplify how societies respond to human needs. Give some examples shown in the video program. Explain why various societies respond differently to human needs.

2. Describe the force driving social change in contemporary societies. Describe and exemplify the impact of technology on society.

3. Distinguish between the social relationships in a *gemeinschaft* and in a *gesellschaft*.

RELATED ACTIVITIES

These activities may be used by your instructor as written assignments or as discussion topics. They may also be included as essay questions on your tests.

1. Describe a role conflict you experience in your life, including the feelings that accompany the stresses involved. In your narrative, identify the two (or more) roles that are in conflict.

2. Role strain is something most of us go through in modern society. Explain how the strain of trying to meet contradictory demands or new expectations in a new role affect you. Identify the role and two or more demands or expectations involved.

3. Often we are not comfortable or happy with our ascribed statuses. Describe ways in which you have had difficulty, discomfort, or unhappiness with one of your ascribed statuses.

4. Using yourself or another person as the example, describe ways in which a master status occasioned prejudice or discrimination. (Master status can be occupational status as well as gender, race, ethnicity, age, or physical appearance.)

5. As an aspect of political socialization, various institutions teach you to respect, love, or identify with your nation. Write a story describing how you or someone you know went through such a socialization experience. Describe the effect of the experience in terms of your perception of the "legitimacy" of the state. (Make up a story if you need to.)

6. One of the basic human needs is the need for food. Imagine that a new breakfast food has been invented, one that is so nourishing and so filling that a midday meal is no longer necessary. What social norms, customs,

institutions, and organization would be affected by this innovation, and how would each change?

PRACTICE TEST

The following items will help you evaluate your understanding of this lesson. Use the answer key at the end of the lesson to check your answers or to locate material related to each question.

Multiple-Choice

Select the one choice that best answers the question.

1. Which of the following best exemplifies a society?
   A. Ireland
   B. Society for the Study of Birds
   C. People residing in the midwestern United States
   D. Environmental Protection Agency

2. A society is a population of people that is characterized solely by its residing in the same geographic region.
   A. subject to similar environmental conditions but lacks social organization.
   B. organized in a cooperative manner to carry out the major functions of life.
   C. required by the nature of its culture to resist outside
   D. interference by violent means through an organized military force.

3. Socially defined positions are known as
   A. groups.
   B. statuses.
   C. societies.
   D. roles.

4. The way a society defines how an individual is to behave in a particular status is referred to as a
   A. role.
   B. status.
   C. social structure.
   D. personality.

5. The combination of a society's expectations about how a role should be performed and the individual's perceptions of what is required is known as
   A. role behavior.
   B. role expectations.
   C. status expectations.
   D. behavioral expectations.

6. The dominant world view of humans of the hunting and gathering period included a
   A. fatalistic acceptance of human frailty in the face of overwhelming natural forces and the need merely to survive.
   B. sense that people can shape their society and should take action to do so.
   C. realization that as society becomes more complex, there is a need for greater specialization.
   D. belief that society should supply a multiplicity of choices, opportunities, and resources.

7. The need to store and defend food surpluses and the formation of villages and small cities are dimensions of what kind of society?
   A. Hunting and gathering
   B. Agrarian
   C. Industrial
   D. Postindustrial

8. Which of the following best describes the changes resulting from the industrial revolution?
   A. Small farms replaced horticultural plots.
   B. People moved from villages to the country.
   C. While eighty percent to ninety percent of the population farmed during the prior 100 years, only three percent to twenty percent were farmers after the industrial revolution.
   D. Immediately after the American Revolution, ninety percent of Americans were farmers; but by the Civil War, only ten percent worked in agriculture.

9. A society based on the close, personal relationships of small groups and communities is referred to as
   A. *gemeinschaft.*
   B. industrial.
   C. secondary.
   D. *gesellschaft.*

10. Yusuf is a police officer. When he arrives home after a particularly harsh day, his two-year-old daughter runs to him and wants his attention and tenderness. He finds it very difficult to "switch gears" emotionally for his daughter.

    Yusuf is experiencing
    A. role containment.
    B. primary relationship syndrome.
    C. role conflict.
    D. role restraint.

11. When people feel anxiety over poor performance of a role, they are experiencing
    A. role strain.
    B. status ascription.
    C. role reversal.
    D. role conflict.

12. Which of the following is an ascribed status?
    A. Waiter
    B. Daughter
    C. Student
    D. Computer programmer

13. Jack is having difficulty finding a job as a nurse. He has discovered that potential employers won't hire him because they believe their patients are accustomed to female nurses.

    Jack is experiencing the results of
    A. role strain.
    B. master status.
    C. role conflict.
    D. status restraint.

14. A society's set of political structures, that is, the structures that deal with questions of "who gets what, when, and how," is known as the
    A. bureaucracy.
    B. country.
    C. state.
    D. community.

15. According to Max Weber, the state gains the right to use force - a key source of its power - from
    A. politicians.
    B. people.
    C. leaders.
    D. divine right.

16. Which of the following is the best example of the results of the lack of match between society and nation?
    A. United States participation in the Vietnam war
    B. United States Constitution
    C. Pledge of Allegiance to the Flag of the United States
    D. Sense of solidarity among American Jews as a people

17. Which statement provides an example of the premise that any societal institution must simultaneously gratify and repress human needs?
    A. Family is an avenue for sexual expression and gratification, as well as control of incest.
    B. Businesses that violate laws regulating interstate commerce can be prosecuted by governmental agencies.
    C. Hospitals provide ways for both doctors and nurses to achieve adequate salaries.
    D. Mass media make millions of dollars every year through the sale of advertising for sports events.

18. The engine or force driving social change in contemporary society is
    A. various cultural ideologies.
    B. opposition to dictatorial regimes.
    C. human anger and aggression.
    D. international competition.

19. Any group in which bonds are in some sense emotional or affective is a
    A. family.
    B. *gemeinschaft*.
    C. society.
    D. *gesellschaft*.

ANSWER KEY

The following provides the answers and references for the practice test questions. Objectives are referenced using the following abbreviations: T = text, R = Telecourse Guide Reading, and V = Video.

Answers	Lesson Goals	Objectives	References
1. A	1	T1	Kornblum, pp. 92-93
2. C	1	T1	Kornblum, p. 92
3. B	1	T2	Kornblum, pp. 93-94
4. A	1	T3	Kornblum, p. 95
5. B	1	T3	Kornblum, p. 95
6. A	2	T4	Kornblum, p. 99
7. B	2	T5	Kornblum, p. 102
8. C	2	T6	Kornblum, p. 106
9. A	3	T7	Kornblum, p. 107
10. C	3	T8	Kornblum, p. 107
11. A	3	T8	Kornblum, p. 108
12. B	3	T9	Kornblum, p. 109
13. B	3	T10	Kornblum, p. 110
14. C	4	T11	Kornblum, p. 110
15. B	4	T12	Kornblum, p. 110
16. D	4	T13	Kornblum, p. 111
17. A	5	V1	Video
18. D	5	V2	Video
19. B	5	V3	Video

Notes and Assignments:

Lesson 9

Cities and Population

LESSON ASSIGNMENT

Review the following assignment in order to schedule your time appropriately. Pay careful attention; the titles and numbers of the textbook chapter, the telecourse guide, and the video program may be different from one another.

Text:

Kornblum, *Sociology in a Changing World*,
Chapter 9, "Population, Urbanization, and Community," pp. 256-283.

Reading:

There is no Reading for this lesson.

Video:

"Cities and Population,"
from the series, *The Sociological Imagination*.

OVERVIEW

On West 42nd Street along lower Times Square, New York City's Public Development Corporation wants to move out the small business owners; the dealers, hustlers, touts, and con men; and the booze, sex, gambling, drugs, and other cheap thrills that form a thriving underground economy and make up a fascinating social world. The authorities want to move in investors and their more upscale businesses by razing the older buildings in the area, thereby supplanting the "less desirable" businesses.

According to sociologist William Kornblum, unless the city addresses the social problems of housing and an unskilled labor force, the "undesirables" on West 42nd Street will simply go somewhere else.

Just twenty blocks northeast of Times Square live Mike and Penny and their five-year-old daughter, Erin. Their apartment is partially subsidized by the hospital down the street, where Mike does his physical therapy research. Tonight the family is giving a dinner party for a relative from the Louisiana delta.

The party will be co-hosted by the neighbor from upstairs and her parents, all of whom are "family" to Mike, Penny, and Erin. These neighbors care for Erin periodically and drop in to talk or have meals together three or four times a week. Of the various friends who will be at the party, several couples have an international mix, such as the Irish-American husband and his Guatemalan wife. Some of the guest are from the neighborhood; some Mike knows from work; and one is an opera singer from Mike and Penny's native Louisiana.

Both of these are vignettes from the life of a major city. And, like much of recent sociology on the subject, they reveal cities as vital human environments where people work, play, and nurture families. Any large city is a mosaic of social worlds and diverse communities. Many cities are centers of culture and finance. But city streets are also the location of the most serious social problems faced by our society.

Urban hospitals oversee thousands of births and deaths every day. People migrate to and from cities in huge numbers each year. A major part of this lesson is about population and how its growth and decline are related to cities and the process of urbanization.

The changes happening in the Times Square and 42nd Street area bespeak both the urban revolution and the conflicts that most North American cities face today. In this lesson you learn about the processes of urban expansion and the changes and conflicts that accompany it.

You also discover that cities are not the faceless, impersonal boiling pot or the sludge of humanity that often is depicted in popular literature and mass media. As with Mike and Penny's family, most urban dwellers form long-term relationships, have an intense sense of community, and think of the city as their home.

This lesson is important because urban centers dominate most cultures and societies throughout the world. A more sophisticated understanding of the many facets of a city will serve you well.

LESSON GOALS

Upon completing this lesson, you should be able to:

1. Analyze the relationship between population changes and urbanization from both historical and international perspectives.

2. Analyze the dynamics of the city over time, including the process of decentralization.

3. Analyze social and economic changes and conflicts among groups in cities.

4. Analyze the diverse ways sociologists and city residents view the city, its community, and its problems.

TEXTBOOK OBJECTIVES

The following textbook objectives are designed to help you get the most from the text. Review them, then read the assignment. You may want to write notes to reinforce what you have learned.

Text: Kornblum, *Sociology in a Changing World*, Chapter 9.

1. Describe demographic transition, and the economic and social developments that are essential if it is to occur.

2. Compare preindustrial cities and modern cities, describing the factors that limited the size of preindustrial cities and the social problems accompanying modern cities.

3. Describe Robert Park and Ernest Burgess's "dynamic model of the city," the process of "invasion," and where immigrant groups move over a long period of time. Be able to recognize examples of invasion.

4. Describe decentralization, and explain how it affects both large and medium-sized cities.

5. Describe the economic changes that have occurred in North America's major cities that are the basis for many urban conflicts.

6. Define gentrification, and discuss how it causes conflicts.

7. Define "defended neighborhoods." Be able to recognize examples.

8. Define gender conflict.

VIDEO OBJECTIVES

The following video objectives are designed to help you get the most from the video segment of this lesson. Review them, then watch the video. You may want to write notes to reinforce what you have learned.

Video: "Cities and Population"

1. Describe the two views of the city that have been held by people, especially throughout the history of the United States, as explained by a sociologist.

2. Describe what the sociologist and the Chicago residents say about the sense of community, connectedness, and belonging among people living or working in the city. Be able to recognize examples.

3. Discuss how people in the underclass experience the city.

4. Identify the problems ordinarily associated with city life. Discuss what sociologist Claude Fischer says about this aspect of living in the city, and what the residents interviewed say about these problems.

RELATED ACTIVITIES

These activities may be used by your instructor as written assignments or as discussion topics. They may also be included as essay questions on your tests.

1. Interview a third-generation immigrant family living in or near a city. Determine where in the city the family members have lived over the three generations. Have they made many moves? Why or why not? Summarize your findings.

2. Describe at least two examples of ways in which residents of your neighborhood or other neighborhoods in your city try to "defend" the neighborhood from "invasion."

3. Were you surprised by the residents' comments about "community in the city"? Describe your reaction to their characterization of community and "belongingness" in the city and why you reacted that way.

4. What is it about a city that brings about a feeling of security and safety? Is it law enforcement; people's sense of community and belonging; physical aspects, such as barriers, parks, and meeting places; or is it something else? Summarize your conclusions.

PRACTICE TEST

The following items will help you evaluate your understanding of this lesson. Use the answer key at the end of the lesson to check your answers or to locate material related to each question.

Multiple-Choice

Select the one choice that best answers the question.

1. The changes in birth and death rates that have occurred primarily in urban industrial nations in the last 200 years are known as
   A. life expectancy.
   B. population amalgamation.
   C. demographic transition.
   D. population intrusion.

2. Preindustrial cities were built
   A. very similarly to the cities we know today.
   B. around a core of retail outlets and markets.
   C. around temples or other ceremonial buildings.
   D. in such a way that the rich lived on the outskirts of the city.

3. Which of the following is NOT one of the factors Robert Park and Ernest Burgess identified as influencing the physical form of cities?
   A. Elevators
   B. Steel construction
   C. High-tech agriculture
   D. Newspapers

4. What is decentralization?
   A. Outlying areas become more important at the expense of the central city.
   B. Governmental institutions become more diversified, with less emphasis on strong federal involvement in urban problems.
   C. Cities expand their boundaries through annexation.
   D. City governments are elected by single-member districts rather than by an at-large system.

5. Which of the following is true of the economic changes that have occurred in North America's major cities?
   A. Many blue collar jobs that once provided employment opportunities for poorly educated residents have either vanished or moved.
   B. There is more competition for the highly prized jobs produced by development of the industrial base.
   C. Numerous educational institutions are training inner-city dwellers for jobs in the industrial and manufacturing sector.
   D. Urban dwellers have moved to the suburbs, where they are participating in a rebirth of manufacturing production.

6. In many poor neighborhoods within the city, wealthier people are buying buildings, forcing the poor to look elsewhere for homes.

   The above is an aspect of
   A. gentrification.
   B. land area reform.
   C. population emergence.
   D. demographic reformulation.

7. Efforts of men trying to "keep women in their place" is called
   A. urban conflict.
   B. racial conflict.
   C. role conflict.
   D. gender conflict.

8. When wealthy suburban neighborhoods pass zoning regulations that establish minimum lot sizes in an attempt to exclude people who cannot afford the large lots, they are known as
   A. invasion suppression groups.
   B. territorial maintenance organizations.
   C. defended neighborhoods.
   D. turf protection units.

9. Which of the following does NOT fit the view of cities described by Claude Fischer?
   A. The city as an association with negative feelings
   B. The city as a preferred place to live
   C. The city as a dangerous, threatening place to live
   D. The city as a place with too much action and entertainment

10. In the introduction of the program, the city is presented as a place where life can be both
   A. harmonious and peaceful.
   B. boring and routine.
   C. overwhelming and creative.
   D. isolated and private.

11. According to Claude Fischer, people in the city are
   A. only loosely connected to the rest of their society.
   B. lonely people in a big crowd.
   C. as socially active and connected as people in a small town.
   D. living fragmented lives that cause them to lead one life with co-workers, one life with neighbors, and one life with friends.

12. According to Sharon Hicks-Bartlett, the underclass in the city
   A. must send their children to schools in the suburbs.
   B. are usually concentrated on the outer edges of the city.
   C. may rarely interact with people outside their communities.
   D. live in areas with more complex social structures than middle class areas have.

13. According to Claude Fischer, which of the following does NOT explain the problem of the city?
    A. Congestion
    B. Bureaucracies
    C. Complexity
    D. Personalities of people

ANSWER KEY

The following provides the answers and references for the practice test questions. Objectives are referenced using the following abbreviations: T = text, R = Telecourse Guide Reading, and V = Video.

Answers	Lesson Goals	Objectives	References
1. C	1	T1	Kornblum, p. 259
2. C	1	T2	Kornblum, p. 264
3. C	2	T3	Kornblum, pp. 267-269
4. A	2	T4	Kornblum, p. 273
5. A	3	T5	Kornblum, p. 279
6. A	3	T6	Kornblum, p. 282
7. D	3	T8	Kornblum, p. 283
8. C	3	T7	Kornblum, p. 282
9. B	4	V1	Video
10. C	4	V1	Video
11. C	4	V2	Video
12. C	4	V3	Video
13. D	4	V4	Video

Notes and Assignments:

Lesson 10

Community

LESSON ASSIGNMENT

Review the following assignment in order to schedule your time appropriately. Pay careful attention; the titles and numbers of the textbook chapter, the telecourse guide, and the video program may be different from one another.

Text:

Kornblum, *Sociology in a Changing World*,
Chapter 9, "Population, Urbanization, and Community," pp. 274-291.

Reading:

There is no Reading for this lesson.

Video:

"Community,"
from the series, *The Sociological Imagination*.

OVERVIEW

It was only about six inches away, but those six inches could just as well have been 600 miles. It was the apartment next door, but Donald and Terri didn't even know the family living there.

With sad smiles on their faces, Donald and Terri tried to joke about their anonymity in this, their residence of two years. Finally, they talked about how alone and isolated they felt—even though they were living in the middle of a city of more than a million people, with one family only the thickness of the wall from them.

"At least we don't have to worry about anyone sticking their noses into our business," Donald tried to console Terri.

“But still, it would be nice to be close to somebody—anybody,” Terri responded. She paused, then tentatively suggested the only idea she had been able to come up with.

“I heard that a neighborhood bar has a friendly social gathering of some families on Sunday afternoons. Maybe we could go and meet some people there.”

“Probably just a bunch of drunks, only interested in football,” said Donald. With his suburban roots, he was skeptical about these urbanites.

But Terri was starting to get excited. “No, one of my co-workers said it's more like a picnic than a drinking party.”

Four months later, Terri and Donald were regulars on Sunday afternoons at Nolan's Saloon and had joined the Tuesday evening bowling league sponsored by the bar. They had become friends and dinner mates with several couples. Finally Terri and Donald were starting to feel like they belonged—as though they were part of something. They even had begun to think of the neighborhood as theirs, not just a place where they had a roof over their heads. They could recognize and greet people on the street. Their lives seemed so much more fun, safer, and warmer now that they were part of a community.

Community occurs in places and situations where you might least expect it. It happens in inner-city neighborhoods and small towns, in professional, recreational, and sacred settings. This lesson explores communities with geographic borders—a territory where people live—but it also considers community as a social structure created by people of similar interests, ethnicity, religion, sexual orientation, or even age.

Community is not just a social structure; it is a quality and a process. We frequently hear words and phrases such as “building a sense of community” or “making a stronger community” or “endangering the spirit of community.”

Since sociologists first began their studies, they've been interested in this fascinating subject. They have approached it in numerous ways and have not always agreed on what community is or how it should be studied. In fact, you will see a great deal of contrast in sociological views about community. There seems to be renewed interest in this topic, and you are fortunate to be in on the ground floor of this new quest of sociological imagination.

In the video program you see and hear several people talk about what it means to them to be part of both territorial and non-territorial communities. Try to find yourself and the people of your “communities” in the pages of the textbook and the frames of the video program.

LESSON GOALS

Upon completing this lesson, you should be able to:

1. Analyze the changing sociological perspectives on the city.
2. Analyze sociological research on the suburban community.
3. Identify recent trends in the development of private communities.
4. Analyze the nature and forms of community, including small and large territorial communities and non-territorial communities.

TEXTBOOK OBJECTIVES

The following textbook objectives are designed to help you get the most from the text. Review them, then read the assignment. You may want to write notes to reinforce what you have learned.

Text: Kornblum, *Sociology in a Changing World*, Chapter 9.

1. Describe sociologists' views of community in the city, and explain where they are contradictory in some instances. Discuss how Robert Park saw the city.
2. Describe the decline-of-community views of the city as developed by Ferdinand Tonnies and Emile Durkheim. Describe what more recent research on bystander apathy has revealed.
3. Describe the more recent sociological view of urban life.
4. Describe the stereotype of suburbs and suburban dwellers held in the 1950s and early 1960s. Describe the conclusions of Herbert Gans about suburbs, and explain how he viewed suburban life compared to central-city life.
5. Identify ways the widespread fear of urban life has altered living styles and communities.

VIDEO OBJECTIVES

The following video objectives are designed to help you get the most from the video segment of this lesson. Review them, then watch the video. You may want to write notes to reinforce what you have learned.

Video: "Community"

1. Describe what communities have provided historically.

2. Identify the advantages and disadvantages of living in a small-town community. Discuss whether territorial communities are on the decline. Provide examples of territorial communities.

3. Explain the problems of the city. Discuss whether or not there is a loss of community in the United States.

4. Identify what people gain from non-territorial communities. Be able to recognize examples evident in the video program on community from the life of the professional woman.

RELATED ACTIVITIES

These activities may be used by your instructor as written assignments or as discussion topics. They may also be included as essay questions on your tests.

1. Write a brief essay or short story in which the main character is a woman who experiences a suburb in the way the research summarized in the textbook section on suburbs portrays the experience.

2. Based on your experience or that of someone you know who has lived in or knows a lot about a small town, describe the advantages and the disadvantages of community in that town. Describe specific instances or situations to illustrate your points.

3. Describe at least one non-territorial community to which you or someone you know belongs. What are the personal and social benefits of belonging to this community?

4. What are three advantages and three disadvantages of living where you do? Name the town or city, and describe the advantages and disadvantages you have identified.

PRACTICE TEST

The following items will help you evaluate your understanding of this lesson. Use the answer key at the end of the lesson to check your answers or to locate material related to each question.

Multiple-Choice

Select the one choice that best answers the question.

1. Social scientists who study life in the cities have devoted a great deal of attention to the tension between "community" and
   A. society.
   B. groups.
   C. socialization.
   D. individualism.

2. Both Ferdinand Tonnies and Emile Durkheim believed that cities
   A. are centers of creativity and individualism.
   B. would soon be replaced by utopian communities.
   C. represent a temporary state in the evolution of human society.
   D. weaken kinship ties and produce impersonal social relationships.

3. Recent research has found that sense of community in urban areas is
   A. often based on the ability to come together by telephone or in special meeting places, such as churches or restaurants.
   B. virtually non-existent due to a decrease in people's sense of belonging within their neighborhood territories.
   C. based on the number of people present in a particular area.
   D. impossible when there is conflict based on ethnicity, religion, or other cultural differences.

4. Which of the following best represents a widely held stereotype of the suburbs and suburban dwellers, especially in the 1950s and early 1960s?
   A. Suburbanites were the picture of spiritual contemplation due to the more pastoral setting of their residences.
   B. The suburbs opposed such cultural activities as theater, music, dance, and art.
   C. The suburbs were the driving force of the economy due to the construction of new homes.
   D. Suburbanites were incapable of real friendships; they were bored, lonely, alienated, atomized, and depersonalized.

5. The growth of private communities is an indication of
   A. the widespread fear of urban life.
   B. the need for expanding populations.
   C. the need for affordable housing.
   D. the need for inner-city development.

6. Throughout the history of North America, our communities have provided
   A. citizens with the scientific knowledge necessary to build an advanced technological economy.
   B. a peaceful, non-violent place to live.
   C. for the basic needs of citizens.
   D. for psychological help from community agencies supported by the state.

7. One of the possible disadvantages of a small-town community is being
   A. located in a rural setting.
   B. subjected to laws that are antiquated.
   C. bored and stifled.
   D. burdened with responsibility.

8. The problems of the city result from
   A. impersonality and a lack of people caring for each other.
   B. neglect by rural legislators who care little about the city.
   C. the structure of society rather than a lack of human caring.
   D. a high degree of mobility among residents.

9. The most common non-territorial community is the
   A. suburb.
   B. political group.
   C. workplace.
   D. family.

ANSWER KEY

The following provides the answers and references for the practice test questions. Objectives are referenced using the following abbreviations: T = text, R = Telecourse Guide Reading, and V = Video.

Answers	Lesson Goals	Objectives	References
1. D	1	T1	Kornblum, p. 275
2. D	1	T2	Kornblum, p. 275
3. A	1	T3	Kornblum, p. 277
4. D	2	T4	Kornblum, p. 278
5. A	3	T5	Kornblum, p. 278
6. C	4	V1	Video
7. C	4	V2	Video
8. C	4	V3	Video
9. C	4	V4	Video

Notes and Assignments:

Lesson 11

Social Control

LESSON ASSIGNMENT

Review the following assignment in order to schedule your time appropriately. Pay careful attention; the titles and numbers of the textbook chapter, the telecourse guide, and the video program may be different from one another.

Text:

Kornblum, *Sociology in a Changing World*,
Chapter 7, "Deviance and Social Control," pp. 190-225.

Reading:

Telecourse Guide for The Sociological Imagination, pp. 134-135.

Video:

"Social Control,"
from the series, *The Sociological Imagination.*

OVERVIEW

Mirra is the eldest and most exuberant of the five children. She is tall for her age, always conspicuous, and the epitome of extroversion. Her mother, Margaret, a devoted and fervent churchwoman, worries about Mirra, especially as the girl reaches puberty.

Delange is a small town. The fear of gossip hangs over town relations like an unseen fog. Within minutes, the gossip network could disseminate emotional and physical details surrounding any of Mirra's actions that might be construed as illicit, illegal, or ill-favored. Margaret dreads the gossip more than death itself, so she continually exhorts her eldest to maintain the highest moral standards and to exhibit the most responsible behavior.

Delange has no formal law enforcement to speak of; its only policeman is drunk most of the time, tolerated solely because of his influential kin. But this community needs no police station. Its centers of social control are the drug store, barber shop, and beauty salon. And how effective they are!

All of us have had to learn the hard way how to control our behavior. We've been punished, perhaps even ridiculed, for socially unacceptable actions. We may have protested the correction, but painful as it was, we know that it probably was necessary. Parents, teachers, family, and friends have—perhaps unwittingly—acted as agents of our society or community in teaching us its rules.

Sometimes we have been rewarded for behaving in accord with the wishes of our group. We received an award, a commendation, praise, or some other sign of approval. These responses are just as surely instruments of social control as are the punishments.

In this lesson you gain greater insight into the meaning and measure of social control. This area of sociological investigation deals with the actions of groups, communities, and societies that function to maintain conformity to social norms and to reduce or eliminate violation of those norms.

Social control is exerted in many different ways. In particular, the video program examines how governments use prisons and other means to try to bring about social control.

You also are encouraged to think critically about these structures and processes and to apply your sociological imagination to one of the most important issues facing communities throughout the United States.

LESSON GOALS

Upon completing this lesson, you should be able to:

1. Analyze social control in its different forms.

2. Analyze institutions of social control, and evaluate the effectiveness of prisons and other strategies as means of social control.

3. Analyze social control in a general sense, and examine prisons and alternatives to imprisonment as institutions of social control.

TEXTBOOK OBJECTIVES

The following textbook objectives are designed to help you get the most from the text. Review them, then read the assignment. You may want to write notes to reinforce what you have learned.

Text: Kornblum, *Sociology in a Changing World*, Chapter 7.

1. Explain the meaning of social control and give examples of it.
2. Determine what sociologists have found in their studies about the equity of, deterrence value of, and trend in public opinion about, capital punishment. Note especially the work of Marvin Wolfgang and Marc Riedel.
3. Identify the functions of prisons according to the ideas of Bruce Jackson.
4. Define a total institution. Cite examples of total institutions and discuss how inmates of total institutions respond to them on both social and personal levels.
5. Describe the conflict and functionalist perspectives on prisons.
6. Discuss the success and failure of the American prison system in rehabilitation of prisoners. Describe the findings of Bruce Jackson and Robert Martinson regarding rehabilitation.

READING
THE MEANING OF SOCIAL CONTROL

READING OBJECTIVES

The following reading objectives are designed to help you get the most from the Reading. Review them, then read the assignment. You may want to write notes to reinforce what you have learned.

Reading: *Telecourse Guide for The Sociological Imagination*, pp. 134-135

1. Describe the different forms of sanctions. Recognize examples of each.

2. Define internalization, and how it relates to social control.

Social Control, Socialization, and Sanctions

Social control refers to all the means that society uses to deal with behavior that violates social norms. These methods vary from a raised eyebrow to capital punishment.

Social control occurs through two processes:

1. the internalization of group norms and
2. the exertion of external pressures in the form of sanctions by others.

The first process, internalization, refers to the kind of learning that takes place when people adopt an attitude or idea as their own. The second process, societal pressure, serves first to teach society's norms and then to remind us of what happens when we deviate from them.

In a sense, much deviance—the violation of social norms—can be seen as a failure in the socialization process. This often results from dysfunctions or inadequacies in the agencies of socialization: church, school, family, mass media. When this happens, the internalization of proper norms, skills, and knowledge does not take place or is defective or inappropriate.

Sometimes people learn norms or attitudes that violate those of the dominant group. Some sociologists see this process as the central cause of the failure of prisons, where people have maximum contact with others who hold "deviant" norms. In this situation, group members are socialized within the inmate subculture, which encourages them to internalize norms, values, and skills that ill-equip them to function as law-abiding citizens in the broader society.

Social control, then, occurs when people internalize the norms of the community. It breaks down when schools, family, and other agencies of the community are unable to do their part in this process. And it is reversed when people learn deviant norms from criminal or delinquent subcultures.

The second aspect of social control, external pressures, are sometimes referred to as sanctions. Sanctions are rewards for conforming to group expectations—and penalties for breaching them. Depending on the social situation, there are four types of sanctions: negative and positive informal sanctions, and negative and positive formal sanctions.

In the Overview, we recognize that the gossip of the Delange barber shops, beauty salon, and drug store occurs in informal chatter among the citizenry. This gossip, therefore, is an informal sanction. Kidding, ridicule, and ostracism are other informal sanctions feared by people in informal groups. Of course, all of these are negative informal sanctions, since having them applied is an unpleasant experience.

However, most of us behave in socially acceptable ways because we have been rewarded by our families and friends for our good behavior. If Mirra is praised for her fine performance in the school dance contest, she is receiving a positive informal sanction.

In social situations governed by formal norms, which are official or written procedures, formal sanctions may be applied. Although in Delange the policeman is a joke and the townspeople in effect police each other, in larger cities the informal ties among citizens may not be as pervasive or as strong. In fact, the sheer size and density of the population make it impossible for people to know everyone else personally. So in urban or even suburban areas there usually is not the same sense of attachment, trust, comfort, and mutual aid that is so pervasive in many very small towns.

Although we have learned that much more community exists in cities than is popularly recognized, cities still require formally organized means of social control, such as law enforcement. If you receive a fine or jail sentence, you have experienced a penalty, which we call a negative formal sanction. On the other hand, if you receive a pay raise, a plaque, or a good grade, you have experienced a positive formal sanction, or reward, for your behavior.

VIDEO OBJECTIVES

The following video objectives are designed to help you get the most from the video segment of this lesson. Review them, then watch the video. You may want to write notes to reinforce what you have learned.

Video: " Social Control"

1. Determine what many sociologists say about the success of prison systems in the United States. Define social control. Explain internalization as it relates to social control.

2. Describe how electronic monitoring works and identify its advantages. Describe the conclusions of the offender under electronic monitoring about this punishment.

3. Define social restitution and explain how social restitution has a positive rehabilitative value, especially as illustrated in the life of the offender working at the hospital.

4. Discuss the benefits of restitution centers described by the offender interviewed at the restitution center and by the corrections expert. Determine what the corrections expert concludes about the community's involvement.

RELATED ACTIVITIES

These activities may be used by your instructor as written assignments or as discussion topics. They may also be included as essay questions on your tests.

1. List five norms or values that you strongly believe in that have helped you to become a participant member of society. Pick norms or values that parents, teachers, or " significant others" instilled in you in your childhood. Explain how beliefs in these norms have contributed to your being an accepted member of your community.

2. Describe two informal positive and two informal negative sanctions used within your family or friendship groups. Describe two formal negative sanctions and two formal positive sanctions that have been applied to you in your life.

3. Have you ever been a participant or inmate in a total institution? If not, think of a book or movie that depicted life in a total institution. From your own experience or from the book or movie, record illustrations of the different aspects of total institutions described in the textbook.

4. What attitudes, practices, norms, or laws exist in your city that do not allow ex-offenders to escape the stigma of their imprisonment? Describe at least three remedies for this situation, that is, three things that could be done to help ex-offenders shed this stigma after they have paid their debt to society.

5. Describe at least two norms that are internalized norms for you. Why do you say they are internalized? What evidence proves your point?

6. What was your reaction to the electronic monitoring and social restitution programs? Do you think these programs offer viable alternatives to prisons? Explain.

PRACTICE TEST

The following items will help you evaluate your understanding of this lesson. Use the answer key at the end of the lesson to check your answers or to locate material related to each question.

Multiple-Choice

Select the one choice that best answers the question.

1. The capacity of a social group, including a whole society, to regulate itself is known as social
   A. regulation.
   B. equilibrium.
   C. control.
   D. boundary maintenance.

2. From comparisons of states with and without the death penalty, evidence on deterrent value has indicated that capital punishment is
   A. effective.
   B. effective in Southern states, but not in Eastern states.
   C. the last, best hope to decrease criminal activity.
   D. ineffective.

3. Which of the following is NOT said to be a function of prisons?
   A. Rehabilitation
   B. Deterrence
   C. Differentiation
   D. Retribution

4. Which of the following does NOT typically occur in a total institution?
   A. Resocialization processes, such as getting haircuts and wearing uniforms, deprive inmates of their former statuses.
   B. Attempts are made to resocialize inmates to behave in ways that suit the organization's needs.
   C. Strong inmate culture results in resistance to officials' control in favor of such values as mutual aid to and loyalty among the inmates.
   D. Inmates are negatively sanctioned for the amount of education they have received and are disgraced by less educated inmates.

5. Functionalist sociologists, such as James Q. Wilson, believe that society should destigmatize crimes and emphasize
   A. rehabilitation.
   B. prisoners should lose their normal citizenship rights.
   C. prisons are schools for crime and should be abolished.
   D. prisoners should receive better rehabilitation but should still be stigmatized in some way.

6. No matter what their perspective, sociologists agree that by far the LEAST successful aspect of prison life is
   A. rehabilitation.
   B. punishment.
   C. restraint.
   D. Retribution.

7. A parent pats a child on the head for behaving nicely during a formal dinner.

   The parent's behavior is an example of what type of sanction?
   A. Negative formal
   B. Negative informal
   C. Positive informal
   D. Positive formal

8. The learning that takes place when people adopt attitudes or ideas as their own is known as
   A. reinforcement.
   B. internalization.
   C. ownership learning.
   D. psychological adoption.

9. Which of the following is a positive sanction?
   A. Being praised
   B. Doing jail time
   C. Getting fined
   D. Being demoted

10. Which of the following is NOT an advantage of electronic monitoring?
    A. Cost is less than that for maintaining the offender in jail.
    B. Cost is paid by the offender, resulting in no expense to the state or city government.
    C. People who do not need prison spend time at home with their families as working, productive citizens.
    D. The offender considers controls never before experienced and subsequently develops internal controls.

11. A means by which an offender can perform community service in an effort to make reparation for the offense that he or she has caused is known as
    A. electronic monitoring.
    B. work release.
    C. social restitution.
    D. probation.

12. Which of the following is NOT mentioned on the video program as an advantage of restitution centers?
    A. They teach individuals to accept responsibility for their behavior.
    B. Life skills programs teach offenders how to maintain employment once it is obtained.
    C. Offenders are taught how to fill out an application, get up on time, and converse with potential employers.
    D. The centers are inexpensive to run because they use abandoned prisons and jails that have been renovated for this minimum security use.

ANSWER KEY

The following provides the answers and references for the practice test questions. Objectives are referenced using the following abbreviations: T = text, R = Telecourse Guide Reading, and V = Video.

Answers	Lesson Goals	Objectives	References
1. C	1	T1	Kornblum, p. 194
2. D	1	T2	Kornblum, p. 211
3. C	2	T3	Kornblum, p. 216
4. D	2	T4	Kornblum, p. 216-217
5. D	2	T5	Kornblum, p. 217
6. A	2	T6	Kornblum, p. 217
7. C	1	R1	TG Reading
8. B	1	R2	TG Reading
9. A	3	V1	Video
10. B	3	V2	Video
11. C	3	V3	Video
12. D	3	V4	Video

Notes and Assignments:

Lesson 12

Deviance

LESSON ASSIGNMENT

Review the following assignment in order to schedule your time appropriately. Pay careful attention; the titles and numbers of the textbook chapter, the telecourse guide, and the video program may be different from one another.

Text:

Kornblum, Sociology in a Changing World,
Chapter 7, "Deviance and Social Control," pp. 190-225

Reading:

Telecourse Guide for The Sociological Imagination, pp. 147-148.

Video:

"Deviance,"
from the series, *The Sociological Imagination.*

OVERVIEW

When Marilyn was busted for the fiftieth time for performing the essential tasks of her profession, she began to wonder where the justice was. She was confused and angry. Her friend Yolanda, less than 250 miles away, operated with little or no harassment from local police. Prostitution was legal in Las Vegas but not in Flagstaff.

The sarcastic police officer had called Marilyn "a deviant." But she remembered reading about how prostitution was accepted in frontier days before the Civil War.

Then she thought about her brother, who had gone to prison back in the late 1960s when vice squad officers had found a joint of marijuana in his car. He was a good kid; now he is stigmatized as an ex-con. Yet these days, possession of small quantities of the drug is legal in his state and only a misdemeanor offense in others.

If what is considered deviant behavior changes so much over time, thought Marilyn, how could it be so wrong?

Marilyn's confusion over the meaning and significance of deviance is shared by many sociologists. We can think of deviance as behavior that strays, or deviates, from the norm of acceptable behavior. But what is acceptable when? Or where? You see in the video program that behavior that comes to be known as deviant occurs in many different social, historical, and cultural contexts.

Also, the label of deviance can be applied to a wide range of behaviors. Some acts that are thought of as deviant by certain segments of society may actually be functional for the society, for example, the creative activities of artists or the social awareness of activists. Other actions that receive little or no public condemnation may be quite dysfunctional and harmful to society, for example, price-fixing and violations of public safety requirements by large corporations.

In this lesson, we explore the meanings of deviance and characteristics of the deviant person. We also see why deviance is studied by sociologists: to help us understand how and why it occurs. The concept of social control is mentioned in the textbook chapter within the context of deviance; however, you already have been introduced to this subject in a previous lesson.

Basically, the first part of this lesson presents the dimensions of deviance, including how it is defined both by various segments of this society and by other, different societies. You also find out about stigma as an aspect of deviance and how definitions of what is deviant change over time. You studied subcultures in a prior lesson; now you learn about deviant subcultures.

Of course, sociologists try to explain social behavior in order to discern what causes it. The lesson introduces you to various theories of deviance, focusing especially on the social-scientific theories, such as functionalist theories, conflict theories, and the explanations of the interactionists.

The Reading centers on elite deviance, filling a gap left by the textbook.

Although deviance is a perplexing reality for sociologists, we must know about it in order to deal with it, because so much of deviant behavior is harmful to society. Most students however, are fascinated by its study. We hope you will be too!

LESSON GOALS

Upon completing this lesson, you should be able to:

1. Analyze deviance as a social, historical, and cultural reality that embraces a wide range of behaviors that can be both functional and dysfunctional for society; analyze the dimensions of deviance, stigma, and deviant subcultures.
2. Analyze the various sociological explanations of deviance, and distinguish between primary and secondary deviance.
3. Analyze the different types of elite deviance.
4. Analyze the social, historical, and cultural contexts of behavior that come to be known as deviant, and examine deviance, particularly elite deviance, in terms of its harm to society.

TEXTBOOK OBJECTIVES

The following textbook objectives are designed to help you get the most from the text. Review them, then read the assignment. You may want to write notes to reinforce what you have learned.

Text: Kornblum, *Sociology in a Changing World*, Chapter 7.

1. Explain and exemplify the meaning of deviance.
2. Define a deviant person and recognize examples.
3. Discuss why certain deviant subcultures—for example, gambling—continue to exist even though there is dishonor attached to them.
4. Differentiate between Goffman's definitions of stigma and of deviance. Recognize examples of stigma and deviance.
5. Explain how variability and change in cultural values and attitudes affect the definition and classification of deviant acts.

6. Explain why Robert K. Merton's theory is referred to as the "anomie" theory of deviance. Recognize examples of deviance caused by anomie.

7. Describe and exemplify the different types of deviance Merton set out in his anomie theory.

8. Describe the Marxian conflict theory of crime.

9. Explain and exemplify both Sutherland's and Cressey's versions of the differential association theory of deviance and criminality.

10. Explain and exemplify the labeling theory of deviance.

11. Discuss what William J. Chamblis discovered about delinquency and the labeling of upper- and lower-class juveniles.

12. Distinguish between primary and secondary deviance. Recognize examples of each.

READING
ELITE DEVIANCE

READING OBJECTIVES

The following reading objectives are designed to help you get the most from the Reading. Review them, then read the assignment. You may want to write notes to reinforce what you have learned.

Reading: *Telecourse Guide for The Sociological Imagination*, pp. 147-148.

1. Define elite deviance.

2. Explain the meaning of and exemplify official deviance.

3. Describe corporate deviance and the extent to which it exists, and explain the unique problems in controlling it.

Elite Deviance

Most people have no difficulty designating armed robbery or murder as deviance, but what about unnecessary surgery a physician performs or the deliberate discharge of dangerous chemicals into the local river by a corporation? Sociologists do not confine their attention and study to the more dramatic and emotionally repugnant forms of deviance. They also delve into the behaviors of individuals, organizations, or institutions whose deviance tends to be shielded from public view and official sanctions.

Elite deviance refers to acts harmful to society that persons from the highest strata of society commit. Some of these acts are criminal; some are unethical, yet not illegal. But, as David Simon and Stanley Eitzen illustrate in their book *Elite Deviance*, all these acts generally are committed with little risk of the perpetrator being apprehended or, if caught, of receiving any severe punishment.

One form of elite deviance is called occupational deviance. This involves illegal and unethical behavior by professional persons, such as doctors, lawyers, and professors. Physicians have easy access to drugs of all kinds and sometimes become addicted to them. A lawyer may become involved in criminal financial manipulation, creating fraudulent tax arrangements for clients, or unethically inflate fees for services performed. Maybe you have known a professor who used his or her authority for sexual harassment. This too is occupational deviance.

Taking advantage of a government position to confer favors or to use public resources for personal gain, to engage in illegal activities or to prevent their detection, illustrates a form of elite deviance known as official deviance. The bribes taken by former vice-president Spiro Agnew and the bribes, hush money, and other maneuvers involved with the Watergate case are examples Craig Little cities in *Understanding Deviance and Control*.

The most harmful form of elite deviance is corporate deviance. A corporation is a large, bureaucratically administered organization whose primary goals are growth and profitability. Corporate deviance is any action committed by an owner or employee of a corporation that violates civil or criminal law.

In *Sociology: A Liberating Perspective*, Alexander Liazos cites the example of a coal company that, despite several warnings, failed to fix a dam. When the dam burst it killed more than 200 people downstream. Liazos also refers to the automobile manufacturer that knowingly produced a car with a gas tank too close to the rear, thus increasing the chance of an explosion after a crash that could burn passengers to death. Liazos comments:

> "It seems obvious that powerful individuals and institutions can kill without their actions being considered murder by law. Indeed, for many years over five times as many people (about one hundred thousand versus twenty thousand) died yearly from accidents at work and diseases they contract at work, than they do from shootings, stabbings, and beatings. But only the latter actions are considered murder. Why?"

Because the corporation is legally treated as an intangible person, says Little, it can be difficult to try to figure out who or what to blame for its misdeeds and then how to punish the miscreant person or institution.

Examples of corporate deviance include knowing and willful violations of occupational health and safety laws resulting in more than 100,000 deaths per year; air, water, and soil pollution in violation of federal standards that contributes to hundreds of thousands of deaths annually; and the twenty million serious injuries each year associated with unsafe and defective consumer products. Corporate deviance is very widespread.

Marshall Clinard, using conservative figures, points out that the majority of the corporations he studied were the object of at least one enforcement action. Large corporations had a disproportionately greater number of violations; the automobile, drug, and oil industries were the most frequent violators. Little reports that eighty percent of them had one or more serious or moderately serious violations, yet eighty percent of the fines imposed were for $5,000 or less. The most frequently imposed sanction was a warning (forty-four percent), followed by monetary penalties (twenty-three percent), and a variety of injunctions or consent orders.

Clearly, the economic power of corporations puts them in the position to avoid prosecution as well as to influence laws that might impede the deviant actions of those corporations.

References:

Liazos, Alexander. 1985. *Sociology: A Liberating Perspective*. Newton, Massachusetts: Allyn and Bacon, Inc., pp. 104-105.

Little, Craig B. 1983. *Understanding Deviance and Control: Theory and Social Policy*. Itasca, Illinois: F.E. Peacock Publishers, Inc., pp. 219-220.

Simon, David R., and D. Stanley Eitzen. 1986. *Elite Deviance*, second edition. Newton, Massachusetts: Allyn and Bacon, Inc., p. 9.

VIDEO OBJECTIVES

The following video objectives are designed to help you get the most from the video segment of this lesson. Review them, then watch the video. You may want to write notes to reinforce what you have learned.

Video: "Deviance"

1. Explain what is meant by the statement: Deviance occurs in a social context. Discuss the kinds of behavior that deviates from a society's norm that can be justified and therefore not be considered deviant by that society. Identify examples of deviance in a social context.

2. Identify examples of deviance in a cultural context.

3. Identify examples of deviance occurring in a historical context.

4. Explain why some sociologists contend that deviance is functional and necessary for society. Identify examples of how deviance can reinforce society's values.

5. Define elite deviance and identify examples. Explain how elite deviance differs from non-elite deviance.

6. Discuss the costs and damage to society of the savings and loan scandal that began in the late 1980s and the sale of the Dalkon shield. Describe the punishment received by those associated with the Dalkon shield.

RELATED ACTIVITIES

These activities may be used by your instructor as written assignments or as discussion topics. They may also be included as essay questions on your tests.

1. Describe the conditions under which you or someone you know (or, if necessary, someone you've heard about) was unjustly labeled as a deviant person. Don't identify the person by name unless the name is already public.

2. Were you ever caught doing something non-conforming or deviant? Did anyone label you a non-conformist or deviant? If so, who? Did the label follow you over a long period of time? Why or why not? Do you think the label became a self-fulfilling prophecy for you? If these questions do not seem to apply to you, find a friend or associate to whom they do apply and ask them the above questions, assuring them of anonymity. Describe their answers using the same labels.

3. Describe a behavior that is considered deviant in one community but acceptable in another. What social institutions or groups dominate the culture of the first community that account for the behavior being defined as deviant?

4. Interview someone from another culture. Ask the person to describe behaviors that culture defines as deviant but that your culture considers acceptable. Report on your findings.

PRACTICE TEST

The following items will help you evaluate your understanding of this lesson. Use the answer key at the end of the lesson to check your answers or to locate material related to each question.

Multiple-Choice

Select the one choice that best answers the question.

1. Byron used inappropriate and explicit sexual references in his conversations at a party and was ostracized by the others there.

   Byron's use of sexual references at the party was
   A. criminal.
   B. deviant.
   C. normal.
   D. illegal.

2. Which action distinguishes a deviant person?
   A. Acting differently from most of the others in a group
   B. Violating laws, no matter how trivial
   C. Opposing a society's most valued norms, especially those that are highly regarded by elite groups
   D. Embarrassing group members to the point of making them feel uncomfortable

3. Which of the following persons is most likely to be classified as deviant?
   A. A person who is involved in stock manipulation
   B. A vagrant who is caught stealing from a convenience store
   C. An artist whose clothing and hairstyle are unconventional
   D. An individual who gets a ticket for jaywalking

4. Certain deviant subcultures, such as prostitution, continue to exist because
   A. they are performing services or supplying products that people in the larger society secretly demand.
   B. there is little or no law enforcement effort to curtail the deviant behaviors of the participants.
   C. ethical values have been eroded by lenient courts.
   D. individuals who participate in them have biological predispositions to act out deviant behaviors.

5. According to Erving Goffman, which term should be reserved for the behavior of "people who are seen as declining voluntarily and openly to accept the social place accorded them, and who act irregularly and somewhat rebelliously in connection with our basic institutions"?
   A. Crime
   B. Deviance
   C. Stigma
   D. Non-conformity

6. "Public perception of the relationship between highway deaths and drunken driving has sharpened as a result of media coverage and lobbying by citizens' groups."

   This statement is an illustration of the
   A. difficulty of classifying deviant behaviors in a rapidly changing society.
   B. way public opinion can change literally overnight.
   C. permanence of certain mores in society.
   D. difficulty of discovering what a society's mores are.

7. It results from the frustration and confusion that people feel when what they have been taught to desire cannot be achieved by the legitimate means available to them.

   "It" refers to
   A. stigma.
   B. frustration-aggression syndrome.
   C. attention-deficit disorder.
   D. anomie.

8. Rickie has given up on his goal to go to college and cares little for the values accepted by most in his society. All that he cares about now is escaping reality by drinking cheap beer all day.

   Using Robert K. Merton's typology, which of the following types of deviance applies to Rickie?
   A. Innovation
   B. Ritualism
   C. Retreatism
   D. Rebellion

9. Legal definitions of deviant behavior usually depend on the ability of the more powerful members of society to impose their will on the government and to protect their actions from legal sanctions.

   This statement represents which approach to deviance?
   A. Anomie
   B. Cultural transmission
   C. Conflict
   D. Differential association

10. Whether a person becomes a criminal is determined largely by the comparative frequency and intimacy of his contacts with law-abiding behavior and criminal behavior.

    This statement best describes which of the following?
    A. Merton's anomie theory
    B. Sutherland's differential association theory
    C. arx's theory of crime
    D. Hirschi's theory of crime

11. Lonnie has lived in a high-crime area most of his sixteen years; most of his close friends are thieves. He sees and does things with them several times every week. Many of the activities of Lonnie and his friends violate the law.

    The description of Lonnie's activities best fits which theory of crime and deviance?
    A. Anomie
    B. Labeling
    C. Marxist
    D. Differential association

12. Deviant behavior is associated with individuals to whom the tag of "deviant" has been successfully applied.

    This statement best represents which theory of deviance?
    A. Labeling
    B. Anomie
    C. Differential association
    D. Conflict

13. In William J. Chambliss's study of the "Saints" and the "Roughnecks" he discovered that members of the
    A. "Saints" were rarely caught and were never labeled as delinquent.
    B. "Roughnecks" were rarely caught and were never labeled as delinquent.
    C. "Saints" and "Roughnecks" were caught at about the same rate and were all labeled as delinquent.
    D. "Saints" were never caught because they avoided truly delinquent behavior.

14. Behavior based on the fact that a person is labeled as delinquent by the authorities is
    A. primary deviance.
    B. anomic deviance.
    C. tertiary deviance.
    D. secondary deviance.

15. Elite deviance refers to harmful actions toward society by individuals
    A. who have been socialized to violence.
    B. opposed to the elites of society.
    C. from the highest strata of society.
    D. striving to gain higher status.

16. Official deviance occurs when a person
    A. uses corporate resources in an illegal manner to make profits for the corporation beyond the bounds of fair trade practices.
    B. who is in government uses public resources for personal gain.
    C. is involved in the acquisition of a position of power through the use of the resources of special interest groups.
    D. helps fellow employees in a corporation cut through bureaucratic procedures in order to gain some personal good.

17. An action violating civil or criminal law committed by owners or employees of a company is
    A. official deviance.
    B. secondary deviance.
    C. corporate deviance.
    D. cultural deviance.

18. Which of the following statements best represents the sociological idea that deviance occurs in a social context?
    A. Deviance is genetically transmitted by one generation to the next.
    B. Deviance is best understood as a result of behavior that violates absolute standards of morality.
    C. Deviance is contained in the interaction between the person, the act, and the people who are responding to it.
    D. Deviance is found in the deviant act itself; that is, deviance is inherently bad.

19. Condoms are typically advertised on Swedish television, but in the United States this practice is not allowed.

    This statement represents which of the following sociological concepts?
    A. Secondary deviance
    B. Cultural context of deviance
    C. Elite deviance in other societies
    D. Historical context of deviance

20. Which of the following best represents the historical relativity of deviance?
    A. Prostitution is legal in parts of Nevada, while in Maine it is illegal.
    B. Twenty-five years ago deviance textbooks listed divorce as a topic; today it is rarely discussed in textbooks.
    C. Murder was a serious crime during the frontier days of American history.
    D. Historically, incest has been a taboo within Western societies.

21. By noting what is deviant, people have a clearer notion of the boundaries of acceptable behavior.

    This statement represents the
    A. function of deviance for society.
    B. intellectual quality of deviance.
    C. historical context of deviance.
    D. role of media in defining deviance.

22. Elite deviance is distinguished from non-elite deviance in that
    A. those engaging in elite deviance are often sociopathic individuals with severe psychological abnormalities.
    B. many of those engaging in it are committing unethical, but not illegal acts.
    C. it is harming society.
    D. elite deviance has insignificant consequences.

23. The device had not been tested in animals or humans for safety, and the manufacturer knew of the inadequacy of testing.

    This statement describes which of the following?
    A. New military equipment development
    B. Development of the most popular treadmill sold
    C. Manufacture of artificial limbs shown in the video program on deviance
    D. Sale and distribution of some products, such as the Dalkon shield, produced by pharmaceutical companies

ANSWER KEY

The following provides the answers and references for the practice test questions. Objectives are referenced using the following abbreviations: T = text, R = Telecourse Guide Reading, and V = Video.

Answers	Lesson Goals	Objectives	References
1. B	1	T1	Kornblum, p. 192
2. C	1	T2	Kornblum, p. 192
3. B	1	T2	Kornblum, p. 192
4. A	1	T3	Kornblum, p. 195
5. B	1	T4	Kornblum, p. 196
6. A	1	T5	Kornblum, p. 198
7. D	2	T6	Kornblum, p. 203
8. C	2	T7	Kornblum, p. 203-204
9. C	2	T8	Kornblum, p. 205
10. B	2	T9	Kornblum, p. 206
11. D	2	T9	Kornblum, p. 206
12. A	2	T10	Kornblum, p. 206
13. A	2	T11	Kornblum, p. 207-208
14. D	2	T12	Kornblum, p. 208
15. C	3	R1	TG Reading
16. B	3	R2	TG Reading
17. C	3	R3	TG Reading
18. C	4	V1	Video
19. B	4	V2	Video
20. B	4	V3	Video
21. A	4	V4	Video
22. B	4	V5	Video
23. D	4	V6	Video

Notes and Assignments:

Lesson 13

Social Stratification

LESSON ASSIGNMENT

Review the following assignment in order to schedule your time appropriately. Pay careful attention; the titles and numbers of the textbook chapter, the telecourse guide, and the video program may be different from one another.

Text:
Kornblum, *Sociology in a Changing World*,
Chapter 11, "Stratification and Social Mobility," pp. 326-357.

Reading:
There is no Reading for this lesson.

Video:
"Social Stratification,"
from the series, *The Sociological Imagination*.

OVERVIEW

The shafts of bright gold in the western sky had given way to a soft purple hue. Ralph sat on a curb, watching the color spectacle and remembering a similar display he'd witnessed with his wife on a family vacation in the West.

But soon he began to feel the cool of the evening seeping through his shirt. He must find something in the trash bin to protect himself from the cold. The now-familiar feelings of anxiety again overtook him. He shuddered, chilled far more by his dilemma than by the night air.

Ralph could hardly believe his plight. Less than a year ago, he had been working at the steel mill before it was shut down. He had tried to find another job, but there was nothing out there for him. He had no family and his marriage was ruined by the stress of unemployment. His friends were sick of his pleas for help.

Ralph was without resource. Demoralized. His life, his hopes and dreams, all had shattered.

Ralph saw a rat sniffing around a trash bin. He felt a strange kinship with the rodent. It was living at the lowest level of existence, surviving off the largess of others. A wistful smile crossed Ralph's face. Tomorrow he would locate the shelter he'd heard about. And he'd try again to find something, anything, to do for honest pay.

Ralph was new to homelessness. He felt confused, dismayed, and powerless over his situation—and profoundly unequal to his former friends and associates, in fact, unequal to just about everyone else in his community.

What Ralph did not understand was that the large European holding company that had purchased the steel mill three years ago had milked it of its profits and invested the earnings in the international banking industry. Economic forces were at work here that had virtually ensured Ralph's downward mobility, his decline in social-class status. He was not the only victim. But by now he felt very much alone.

In this lesson you see how systems of social stratification—the unequal distribution of the socially valued rewards across a community or society—have developed historically and internationally. You find out that people's acceptance of their "place" in society is based not just on their individual personalities but on social forces, such as culture and socialization.

By watching the video program and reading the textbook, you explore the meaning and dimensions of social stratification, how it exists in a small town and other societies. You encounter a woman who is on the edge of downward mobility. You see and hear explanations that will help you to understand Ralph's predicament. What type of mobility did he experience? What economic, social, and historical forces were at work to place him alone on that curb at the margins of the city?

Most Americans prefer to believe that equality is the reality in their society. Since they are uncomfortable and confused by the many reminders of the stratified nature of their communities, they would rather ignore them. This lesson gives you perspective and new insights into the difficult and often troublesome reality of social stratification.

LESSON GOALS

Upon completing this lesson, you should be able to:

1. Analyze social stratification and inequality at the macro and micro levels, and how mobility varies in different stratification systems.
2. Analyze how social stratification changed with the industrial revolution.
3. Analyze the views of Karl Marx and Max Weber about class and class conflict.
4. Analyze the conflict, functionalist, and interactionist theories of social stratification.
5. Analyze the stratified nature of social structure, and examine the different types of stratification systems throughout history.

TEXTBOOK OBJECTIVES

The following textbook objectives are designed to help you get the most from the text. Review them, then read the assignment. You may want to write notes to reinforce what you have learned.

Text: Kornblum, *Sociology in a Changing World*, Chapter 11.

1. Differentiate between inequality and social stratification. Include how widespread inequality is.
2. Define each of the following, and explain what each has to do with social mobility: open and closed stratification systems, castes, ascribed and achieved statuses, and class.
3. Determine what the main factor is in determining our positions in the stratification system. Define structural and spatial mobility.
4. Explain and give examples of the following: status symbols, deference, and demeanor.

5. Explain the difference between power and authority. Include examples of each.
6. Describe the industrial revolution, its historical development in different parts of the world, and the technological innovations that accompanied it.
7. Identify the important innovations that accompanied the industrial revolution.
8. Discuss how Karl Marx defined classes. Explain what he concluded about class conflict.
9. Explain the meaning of class consciousness. Recognize examples of class consciousness.
10. Discuss how Max Weber defined social class. Recognize examples illustrating his point.
11. Explain and give examples of intragenerational mobility and intergenerational mobility.
12. Explain various versions of the conflict approach to social stratification. Discuss whether or not Karl Marx's predictions have occurred.
13. Identify the functionalist explanation of social stratification. Discuss what critics say about this explanation. Cite examples used in this criticism.
14. Determine what interactionist sociologists focus on in their approach to social stratification.

VIDEO OBJECTIVES

The following video objectives are designed to help you get the most from the video segment of this lesson. Review them, then watch the video. You may want to write notes to reinforce what you have learned.

Video: "Social Stratification"

1. Determine the most rigid form of stratification system. Describe this system, as well as the type of social stratification that existed in many agrarian societies, such as China before the revolution. Be able to recognize examples.

2. Discuss how the market system and the industrial revolution were related to each other. Explain their effects both on people and on social organization.

3. Determine the conditions that social stratification systems develop. Explain the arguments about whether social stratification is essential. Discuss how the rapid developments in science and technology have changed social organization and culture.

4. Determine what single factor has been most important in creating patterns of downward mobility in recent years. Describe the downward mobility of some American women, using the woman in the video program as an example. Identify which segment of society is growing richer and which segment is becoming poorer since the 1960s and explain why.

RELATED ACTIVITIES

These activities may be used by your instructor as written assignments or as discussion topics. They may also be included as essay questions on your tests.

1. Pretend that you have won a two-million-dollar lottery. Let us assume that you wish to use the money to become upwardly mobile. How would you spend it? Would your choices be based on your present tastes? If not, how would you acquire "higher" tastes? How would your relationships with friends change? What would be the psychological and social costs of this new life to you and your family?

2. Are you class conscious? Do you accept your class as right, proper, and equitable? Explain your answers.

3. Describe an individual in your community who has a lot of prestige or power or both but not much wealth. What social class would you put that person in? Explain how he or she illustrates either Karl Marx's or Max Weber's view of social stratification and mobility.

4. Do you know of a divorced woman who has experienced downward mobility? Describe the evidence of her downward mobility.

PRACTICE TEST

The following items will help you evaluate your understanding of this lesson. Use the answer key at the end of the lesson to check your answers or to locate material related to each question.

Multiple-Choice

Select the one choice that best answers the question.

1. A society's system for ranking people according to attributes such as income, wealth, power, prestige, age, sex, ethnicity, or religion is called social
   A. inequality.
   B. caste.
   C. structure.
   D. stratification.

2. A status that is acquired at birth is known as
   A. ascribed.
   B. achieved.
   C. open.
   D. dominant.

3. The main factor determining position in our society's stratification system is our
   A. relationship to the means of existence.
   B. religious affiliation.
   C. intelligence.
   D. personality characteristics, such as aggressiveness.

4. Jane holds the door open for her teacher, smiling and greeting her with the words: "Good morning, Professor Kelly. How is the professor this morning?"

   Jane has just practiced
   A. ideology.
   B. legitimacy.
   C. deference.
   D. demeanor.

5. Oscar forces all the children in the block to bring him candy by threatening to hurt them.

   This is an example of the exercise of
   A. authority.
   B. authority but not power.
   C. cultural ambivalence.
   D. power.

6. The industrial revolution
   A. is still occurring in China, India, and Africa.
   B. was the basis for ending slavery in the United States.
   C. began in the former colonial outposts of England, France, and Germany.
   D. witnessed an increasing proportion of the population involved in farming as opposed to entrepreneurship.

7. Which of the following was NOT an important innovation in social institutions accompanying the industrial revolution?
   A. Demands for full political rights
   B. Sales of labor for wages in factories and commercial firms
   C. Relationships based on contracts
   D. Relationships based on reciprocal obligations

8. Marx argued that business competition would eventually result in
   A. monopolies through elimination of less successful firms.
   B. greater productivity among economic units.
   C. the proletariat becoming the dominant force in society.
   D. increasing complexity among cultures.

9. The workers at an industrial plant have risen up in collective opposition. They have made a statement that the management and owners of the company are exploiting them by extracting too much profit for stockholders and by not investing in new equipment and fair salaries for workers.

   This scenario in the industrial plant is an example of workers
   A. being an objective, but not a subjective, class.
   B. having class consciousness.
   C. expanding the objective class.
   D. avoiding establishment of a subjective class.

10. Max Weber defined social class in terms of
   A. purely economic factors.
   B. wealth, prestige, and power.
   C. Karl Marx's philosophy.
   D. possession of power only.

11. Dominique's parents were working class people. She has become a physician.

   Dominique is an example of
   A. structural mobility.
   B. intragenerational mobility.
   C. intergenerational mobility.
   D. group mobility.

12. Which of the following is NOT an accurate characterization of Karl Marx's conflict perspective?
    A. Capitalist societies are divided into two opposing classes, wage workers and capitalists.
    B. Social classes emerge because an unequal distribution of rewards is essential in complex societies.
    C. Conflicts between the classes will eventually lead to revolution.
    D. Revolutions of workers will result in the establishment of classless, socialist societies.

13. Which theory embodies the concept that social stratification is necessary if society is to match the most talented individuals with the most challenging positions?
    A. Functionalist
    B. Conflict
    C. Interactionist
    D. Ecological

14. Within economic classes, people form status groups whose prestige is measured not according to production or wealth but according to what they communicate about themselves through their purchases.

    The above is a statement best representing which theory of social stratification?
    A. Functionalist
    B. Interactionist
    C. Conflict
    D. Ecological

15. The most rigid form of social stratification system is a
    A. class system.
    B. caste system.
    C. socialist system.
    D. *gemeinschaft* system.

16. The change in production from the use of human muscle power to the use of machinery and inanimate sources of energy is called the
    A. technological transition.
    B. productive transformation.
    C. industrial revolution.
    D. dynamic of the caste system.

17. Social stratification systems develop when
    A. division of labor is minimal.
    B. people acquire more valuable resources.
    C. populations migrate from one area to another.
    D. animals available for hunting decrease in number.

18. Women experience downward mobility because they
    A. were abused as children.
    B. are competent but unwilling to seek promotions.
    C. get divorces and are no longer able to make house payments.
    D. cannot stand insults and anger often encountered in the business world.

ANSWER KEY

The following provides the answers and references for the practice test questions. Objectives are referenced using the following abbreviations: T = text, R = Telecourse Guide Reading, and V = Video.

Answers	Lesson Goals	Objectives	References
1. D	1	T1	Kornblum, p. 328
2. A	1	T2	Kornblum, p. 328
3. A	1	T3	Kornblum, p. 337
4. C	1	T4	Kornblum, pp. 338-339
5. D	2	T5	Kornblum, pp. 339-340
6. A	2	T6	Kornblum, p. 342
7. D	2	T7	Kornblum, p. 343
8. A	3	T8	Kornblum, p. 345
9. B	3	T9	Kornblum, p. 345
10. B	3	T10	Kornblum, p. 346
11. C	3	T11	Kornblum, p. 346
12. B	4	T12	Kornblum, p. 348
13. A	4	T13	Kornblum, p. 349
14. B	4	T14	Kornblum, p. 349
15. B	5	V1	Video
16. C	5	V2	Video
17. B	5	V3	Video
18. C	5	V4	Video

Notes and Assignments:

Lesson 14

Social Class

LESSON ASSIGNMENT

Review the following assignment in order to schedule your time appropriately. Pay careful attention; the titles and numbers of the textbook chapter, the telecourse guide, and the video program may be different from one another.

Text:

Kornblum, *Sociology in a Changing World*,
Chapter 12, "Inequalities of Social Class," pp. 358-395.

Reading:

There is no Reading for this lesson.

Video:

"Social Class,"
from the series, *The Sociological Imagination*.

OVERVIEW

These were the first clients Darrell would visit as a new social-work student. As he parked his car, he glanced again at the dilapidated old house. A patched-up fence around the tiny yard acted as an enclosure for numerous farm animals. Their presence, against city ordinances, was signaled by the stench that invaded Darrell's nostrils as he climbed out of his car. The novice social worker's emotions were running high. Was he ready for this?

The screen door still had the huge rip in it noted by the previous welfare worker. Darrell knocked on the door frame and heard a faint voice respond, "Come in." The wife, pregnant again, was rocking on one of the lawn chairs that served as the living room's furniture. The husband, whose chief employment was doing odd jobs, painting and doing some fishing, acted shy and somewhat tongue-tied. The children seemed lethargic; their eyes, sad and dull.

After introducing himself and attempting to converse with the couple, Darrell asked if he could take a look in their kitchen. He needed to determine whether the family's nutritional habits had improved any, since the social-work professionals were concerned about the ability of the parents to care properly for their children.

But the refrigerator was nearly empty; its freezer was entirely iced over. The cupboards held a few cans, which wouldn't feed Darrell's middle-class family for a day. As he walked back past the one bedroom, he saw flies buzzing about the baby, who lay apathetically on the old bed.

Darrell had read about conditions like these and had heard other students swapping stories. But, until now, he had never seen poverty firsthand—up close and personal. He felt both sad and horrified, with an uncertain, nagging fear. How could people be living like this—in the midst of the same city he had grown up in?

One of the very real consequences of the social-stratification system that exists in the United States is poverty. In this lesson we explore the dimensions and results of America's social-class system, raising some very pointed questions. What are the class divisions? How are they measured? How have perspectives on social inequality changed over time? What is the extent of upward social mobility, and how does a person achieve it? And how does poverty fit into the social-class picture?

The video program introduces you to two young women and their families. You take a peek into their lives and have a chance to try to understand how their customs and mentality affect their opportunities. If you look closely, using your sociological imagination, you also see how their lifestyles manifest the social-class divisions of their communities. This sociological interpretation helps us better understand both the limits and the opportunities inherent in America's social-class system.

Many Americans are reluctant to admit that inequality exists in this society, that the cultural ideal of equal opportunity often is more myth than reality. But the facts about our social-class structure give you all the more reason to develop your sociological imagination about this fundamental aspect of social reality.

LESSON GOALS

Upon completing this lesson, you should be able to:

1. Analyze how sociologists measure social inequality.

2. Analyze the makeup of the different social classes and what "life chances" each has.

3. Analyze the magnitude of poverty and who the poor are in the United States.

4. Analyze social mobility and the functions and dysfunctions of poverty.

5. Analyze social class in the United States by focusing on two teenage girls from different classes, and then analyze whether the social-class system in America is discriminatory.

TEXTBOOK OBJECTIVES

The following textbook objectives are designed to help you get the most from the text. Review them, then read the assignment. You may want to write notes to reinforce what you have learned.

Text: Kornblum, *Sociology in a Changing World*, Chapter 12.

1. Identify the basic measures of social inequality in the United States. Explain how social inequality is distributed in the United States.

2. Identify the major sociological studies of social class in American communities in the 1930s and 1940s. Describe the most important findings of each. NOTE: These studies are usually identified by the names of the sociologists doing them.

3. Describe the changes in social classes and in class lines that have occurred with the coming of the postindustrial society.

4. Describe how social class is related to health, education, and politics.

5. Identify the percentage of the U.S. population that comprises the upper class. Identify how much of the personal wealth of the United States they control. Describe the types of people who are in the upper class.

6. Describe the elitist and pluralism models of class.

7. Distinguish between the middle class and the upper-middle class. Describe the size and diverse qualities of the middle class.

8. Describe the changes taking place in the working class. Explain the divisions and diversity in the working class.

9. Distinguish between absolute and relative deprivation and be able to recognize examples of each.

10. Explain why we have increasing poverty amid affluence.

11. Explain what is meant by "the feminization of poverty."

12. Explain Herbert Gans's conclusions about the functions and dysfunctions of poverty.

13. Identify where the poor live. Explain the relationship between poverty and place.

VIDEO OBJECTIVES

The following video objectives are designed to help you get the most from the video segment of this lesson. Review them, then watch the video. You may want to write notes to reinforce what you have learned.

Video: "Social Class"

1. Describe the gradational concept of social class. Explain what determines the placement of people in a social class. Give examples.

2. Explain the social-relations concept of social class. Be able to recognize examples.

3. Describe the differences in cultural practices of various classes.

4. Explain the importance of education in terms of social class.

5. Identify the discriminatory aspects of social class. Explain how these aspects should be dealt with.

RELATED ACTIVITIES

These activities may be used by your instructor as written assignments or as discussion topics. They may also be included as essay questions on your tests.

1. Describe the divisions that exist in your community between the working class and the upper class. What evidence illustrates the gap in income and lifestyle between the two classes?

2. In which social class do you fit? (You may use any of the descriptions of social classes described in the textbook, but mention which one you are using.) Are there aspects of your lifestyle, income, or job that would place you in two social classes?

3. How is your education contributing to your social-class standing now and in the future? How is it likely to affect your health and other aspects of your "life chances"?

4. Do you know a person in the working class who illustrates some of the changes described in the textbook? Into which of the two major divisions of the American working class does or did the person fit?

5. When you were a child, did your parents or others put pressure on you to perform well in school? If so or if not, how did this affect your behavior, attitude, values, and other responses?

6. Which of the two young women featured in the video program on social class did you identify with more? Why?

7. Has your social-class position constrained or helped you in achieving your educational goals? Explain.

PRACTICE TEST

The following items will help you evaluate your understanding of this lesson. Use the answer key at the end of the lesson to check your answers or to locate material related to each question.

Multiple-Choice

Select the one choice that best answers the question.

1. Which of the following statements is true regarding the distribution of social inequality in the United States?
   A. Forty percent of American households have a net worth of $250,000 or more.
   B. Distribution of wealth and income is more equal than distribution of occupational prestige and educational attainment.
   C. The prestige people attach to an occupation is heavily influenced by the education required for the job.
   D. Educational attainment has become less equal over time in America.

2. *Middletown* provided numerous examples of the
   A. gap in income and lifestyle between the working class and the owners or managers of capital.
   B. prestige patterns typical of most American cities with a population of at least 100,000.
   C. pattern of residential segregation that typically occurred in large cities, such as Chicago.
   D. racial caste system of the South.

3. Which of the following is a result of the coming of the postindustrial society?
   A. Clearer delineation of the classes
   B. Formation of three distinct social classes
   C. Blurring of class lines
   D. Formation of a techno-class

4. Which class is less likely to be exposed to toxic chemicals or to experience occupational stress and peptic ulcers?
   A. Lower
   B. Middle
   C. Upper
   D. Poor

5. The public schools that serve the middle classes, compared to those that serve the working class,
   A. spend more per pupil and offer a wider array of special services.
   B. enroll students with higher intelligence and are therefore more effective as college prep schools.
   C. end to have bureaucracies that are more complex and have a higher percentage of administrators.
   D. have higher student-teacher ratios.

6. Which of the following is the most accurate representation of the upper classes?
   A. They tend to be from Catholic and Jewish families.
   B. They control five percent of all personal wealth in the United States.
   C. They create special places in which to live and relax.
   D. They send their children to public universities.

7. Those who support the elitist model of social class argue that the ruling class
   A. commands much prestige but has a moderate amount of power.
   B. holds a virtual monopoly over power in the united states.
   C. possesses power but lacks unity in its execution.
   D. pulls the purse strings but does not have political power in the traditional sense.

8. Which of the following is true of the upper-middle and the middle classes?
   A. They can be easily differentiated.
   B. They comprise about seventy percent of the population.
   C. They tend to be employed in non-manual occupations.
   D. They gain status through inherited wealth.

9. Which social class is undergoing the most rapid and difficult changes in America today?
   A. Upper
   B. Middle
   C. Upper-middle
   D. Working

10. Poor families in Africa or India may be living without permanent shelter and be forced to beg and sift through scraps for enough food to survive.

    This description is an expression of
    A. absolute deprivation.
    B. working class deprivation.
    C. relative deprivation.
    D. social change in third world nations.

11. A major reason we have increasing poverty amid increasing affluence is due to the
    A. growing number of single-parent, female-headed families.
    B. increasing complexity of our society.
    C. transition to a more production and manufacturing oriented economy.
    D. lack of adaptability of colleges and universities in responding to the poor as a social class.

12. Of the poor in the United States,
    A. more than fifty percent are concentrated in the large central cities.
    B. most live in the high-poverty neighborhoods of the central cities.
    C. twenty-nine percent reside in rural and small-town communities.
    D. fewer than one percent live in the suburbs of large cities.

13. Which of the following is listed by Herbert Gans as a function of poverty?
    A. It decreases population pressures on the society.
    B. It allows the economy to diversify.
    C. It is an alternative to suicide.
    D. It creates jobs for professionals, such as police and social service workers.

14. In comparison to affluent neighborhoods, “poor places” typically offer
    A. poorer schooling.
    B. more recreational opportunities.
    C. increasing entry-level job prospects.
    D. lower rates of substance abuse.

15. The “gradational concept” of social class is a concept in which people from the top and the bottom are differentiated by their
    A. prestige.
    B. income.
    C. personality.
    D. influence.

16. Which of the following best represents the social relations concept of social class?
    A. Relationships differentiated as upper, upper-middle, middle, working, and lower classes
    B. Relationships across the six classes typically listed by sociologists
    C. Relationships within the working class
    D. Relationships between capitalists and workers

17. The single most important determinant of an individual’s place in the class system centers around the
    A. amount of prestige the person has.
    B. development of a positive attitude.
    C. ability to keep informal ties and friendships.
    D. kind of educational qualifications that the person acquires.

18. According to Professor Erik Olin Wright, interviewed in the video program on social class, the social class system in America
    A. functions properly in a complex economy.
    B. encourages achievement.
    C. causes discrimination.
    D. provides more rewards to those who work harder.

ANSWER KEY

The following provides the answers and references for the practice test questions. Objectives are referenced using the following abbreviations: T = text, R = Telecourse Guide Reading, and V = Video.

Answers	Lesson Goals	Objectives	References
1. C	1	T1	Kornblum, p. 365
2. A	1	T2	Kornblum, p. 366
3. C	1	T3	Kornblum, p. 371
4. C	2	T4	Kornblum, p. 373
5. A	2	T4	Kornblum, p. 373
6. C	2	T5	Kornblum, p. 376
7. B	2	T6	Kornblum, p. 376
8. C	2	T7	Kornblum, p. 377
9. D	2	T8	Kornblum, p. 378
10. A	3	T9	Kornblum, p. 379
11. A	3	T10	Kornblum, p. 380
12. C	3	T13	Kornblum, p. 381
13. D	4	T12	Kornblum, p. 385
14. D	3	T13	Kornblum, p. 385
15. B	5	V1	Video
16. D	5	V2	Video
17. D	5	V4	Video
18. C	5	V5	Video

Notes and Assignments:

Lesson 15

Race and Ethnicity

LESSON ASSIGNMENT

Review the following assignment in order to schedule your time appropriately. Pay careful attention; the titles and numbers of the textbook chapter, the telecourse guide, and the video program may be different from one another.

Text:

Kornblum, Sociology in a Changing World,
Chapter 13, "Inequalities of Race and Ethnicity," pp. 396-435.

Reading:

Telecourse Guide for The Sociological Imagination, pp. 190-191.

Video:

"Race and Ethnicity,"
from the series, *The Sociological Imagination*.

OVERVIEW

His eyes seemed to glaze over as if he were mentally in another place. Gloria had seen that look before. But she had believed that looking for a job would be different now: She finally had her college degree. But that certainly wasn't going to change anything here. "We just don't have anything now, little lady," Mr. Glazed Eyes said.

Gloria's reaction to the eyes was fear; to "little lady," she flashed with anger. But she did not betray her feelings. In her native Spanish her *madre* had told her, "Don't give them the satisfaction," advising a psychological protection against prejudice.

Gloria's Anglo friend in the small company had told her of the opening in accounting, so she knew Mr. Glazed Eyes was lying. Now she had a decision to make. Should she pursue this thing with legal action against the obvious discrimination, or should she let it lie?

Her *abuelo* (grandfather) would have advised her not to make trouble. But in an instant she was back in the office of her *tio* (uncle), director of the El Paso clinic nestled in the poorest neighborhood in the country. In her mind's eye she saw—proudly displayed on two walls—the photos of Cesar Chavez, their people, and other leaders and fighters for equality. In their eyes she read determination; in the lines etched deeply into their faces, the sadness wrought from a life of oppression and exploitation.

Gloria recalled how her *tio* would tease visitors who asked him which tribe was symbolized by the turquoise locket hanging from his neck. With just a hint of a grin, he would proclaim, "It is the emblem of the Chicano Nation."

As these visages replaced the face of Mr. Glazed Eyes, Gloria knew what she must do.

This lesson helps you to understand the sociological story behind Gloria's immediate dilemma. Why did so many Anglos assume that Gloria's uncle was an American Indian? Was it his facial features or the necklace? Was her uncle sending a sociological message of ethnic nationalism and pride in his quip about the "Chicano Nation"? Did Anglos respond to his different racial or his ethnic (cultural) characteristics? What is race, anyway?

But Gloria's immediate concern was much easier to define. She had worked hard to earn her college degree; she met all the qualifications the Anglo world had set for the accounting position. And she had been hit—once again—with a response over which she had no control, a response that bore no relationship to her credentials, capabilities, personality, or personal dignity.

Gloria was aware of the history of minority-majority relations in North America, laced with assorted strategies for maintaining power—from genocide to forced assimilation. She also knew of the range of possible responses to her situation—from acceptance and acquiescence to legal contest to collective confrontation.

But the question we all must ask is: What was really behind the potential employer's refusal? Prejudice or discrimination? And what is the difference between the two?

As uncomfortable as it might be to face these realities, understanding is essential to dealing effectively with the social problems that accompany racial and ethnic inequality. You too will become a more informed and competent citizen because of your journey into this arena of sociological investigation.

LESSON GOALS

Upon completing this lesson, you should be able to:

1. Analyze which categories of people fit the sociological definition of a minority group, and analyze the significance of race, racism, and ethnicity.

2. Analyze the various ways the dominant group has treated minorities throughout the history of the United States.

3. Analyze the dynamics of prejudice and discrimination, and analyze various responses to them.

4. Analyze theories of race and ethnic inequality.

5. Analyze the social problems related to race and ethnic inequality.

6. Analyze prejudice and discrimination and how they affect members of various ethnic groups.

TEXTBOOK OBJECTIVES

The following textbook objectives are designed to help you get the most from the text. Review them, then read the assignment. You may want to write notes to reinforce what you have learned.

Text: Kornblum, Sociology in a Changing World, Chapter 13.

1. Define racism, and explain why attempts to show racial differences in I.Q. have failed.
2. Explain the meaning of minority groups, and explain why a minority is not necessarily numerically smaller than the dominant group.
3. Define and give examples of genocide.
4. Describe expulsion, and cite several historical instances in which expulsion has occurred in the United States.
5. Explain the meaning of slavery, and describe the magnitude of the transatlantic slave trade.
6. Define segregation and identify the different types of segregation. Explain how segregation has been dealt with by the U.S. Supreme Court and Congress.
7. Define assimilation as it relates to intergroup relations, and describe how it has occurred in Latin America.
8. Describe a pluralistic society, and explain why is it difficult to achieve.
9. Explain the meaning of stereotypes, and recognize examples.
10. Differentiate between prejudice and discrimination, recognize examples of each, and explain Tobert Merton's typology of the two.
11. Identify the arguments for and against affirmative action. Be able to recognize examples.

12. Explain the frustration-aggression hypothesis; describe the characteristics of the authoritarian personality, and be able to recognize examples based on these theories.

13. Discuss how the interactionists explain intergroup hostility.

14. Describe and give examples of the functionalist explanation of inequality among racial and ethnic groups. Explain the limitations of the theory in trying to understand both problems and solutions relative to apartheid in South Africa.

15. Describe Karl Marx's theory of racial and ethnic inequality.

16. Explain the conclusions of William Julius Wilson and Stanley Lieberson about the problems of lagging black mobility. Be able to recognize examples.

READING
PREJUDICE AND DISCRIMINATION AND RESPONSES TO THEM

READING OBJECTIVE

The following reading objective is designed to help you get the most from the Reading. Review it, then read the assignment. You may want to write notes to reinforce what you have learned.

Reading: *Telecourse Guide for The Sociological Imagination*, pp. 190-191.

1. Explain prejudice as an attitude, including its three components. Be able to recognize examples.

Prejudice as an Attitude

Attitudes are to our personalities what rivers are to the land. Along and through our personalities flow the attitudes that feed our behavior. We have attitudes about many different subjects and issues: sex, religion, family, friends, race, computer technology, school.

For example, assuming that you have learned a positive attitude toward formal education, the ideas you learn during the educational process stimulate you and fill your mind. You consider and cherish the challenges and issues you discover in school. You feel excited about school and attracted to it. Educational success gives you joy and fulfillment.

As a result, your behavior reflects your attitude. You tend to spend your time in educational pursuits. They become the subject of your conversations. You invest money in them. All of this indicates your attitude toward education.

An *attitude* is a predisposition to think, feel, and act in a certain way. Because many of our attitudes are so deeply ingrained in our personalities, three is a tendency to think of them as somehow inherent in our being. Attitudes are not inborn; they are learned. You were not born with your attitudes toward education; you learned to like school.

So it is with prejudice. As the song from the old musical *South Pacific* said, "You have to be taught to hate and fear, to hate all the people your relatives hate." Since these attitudes are learned, they have to be acquired from somewhere. That somewhere is the culture and the minds of those who carry the culture to us. That includes all the agencies of socialization: family, friends, mass media, religion, and school. Thus, prejudice is a sociocultural phenomenon. It emerges from groups and affects interpersonal relations for the worse. It causes suspicion, discord, violence, discrimination, and other social problems.

In the context of this lesson, *prejudice* is a negative attitude toward some group, a prejudging of a person who allegedly belongs to a group you have negative feelings about, before you get to know the person as an individual. Prejudice is learned by interacting with others who carry similar attitudes. A *stereotype* is an aspect of prejudice. It is an overgeneralized—and therefore incomplete and inaccurate—image of a whole category of people. Seeing women as emotional and irrational, blacks as lazy, and Latinos as violent are examples. Although there can be a grain of truth in stereotypes, the thinking based on these easy, simplistic categories is based on ignorance, an incomplete knowledge of the judged person or group. This type of thinking can be comforting for the insecure individual. In this sense, Archie Bunker's ignorance was his bliss.

All our prejudices involve negative feelings or emotions, such as hostility, fear, envy, pity, guilt, aggression, and anger. Although our prejudices may emerge as a result of our thinking those negative thoughts and accepting stereotypes as facts, these feelings are often unconscious and therefore difficult to control. They pop up spontaneously and involuntarily. Therefore, the "feeling" components of our prejudices often linger long after we've done away with believing stereotypes intellectually.

Unfortunately, when we think or feel in a prejudiced manner, we tend to act in that manner. For example, if a top manager thinks "women are irrational," he may be inclined to pass over any woman when an opportunity for promotion comes around. Some white neighbors' irrational fears of black people may cause the whites to think about putting security bars on the windows of their homes in areas where crime is not a problem, or even to sell their homes to "block busters" for a pittance, when one black family buys a home in the neighborhood.

A tendency to discriminate is an aspect of prejudice, although discrimination itself is an action, not a feeling. Discrimination involves denial of various kinds of experiences to people based on their allegedly belonging to a specific racial or ethnic—or any other—category. If that top manager follows through on his prejudice toward all women and denies the qualified woman her promotion, he has discriminated against her.

In sum, *prejudice* is a learned attitude that involves negative thoughts (stereotypes), feelings (such as hostility or fear) and the tendency to act in a discriminatory way. Prejudice is learned; therefore it is part of the culture and the society in which it exists.

Many people claim that American society is color conscious. Because prejudice permeates the sociocultural fabric, most of us have some degree of prejudice toward some group or other—whether we ourselves are part of a majority group, or a minority group. However, if we are to reduce our prejudices, it is important for us to understand their dynamics. As long as our prejudices remain totally at the unconscious level, like Archie Bunker's, they control our thinking and feeling—without giving our minds a chance to let us see the beauty of a rainbow.

VIDEO OBJECTIVES

The following video objectives are designed to help you get the most from the video segment of this lesson. Review them, then watch the video. You may want to write notes to reinforce what you have learned.

Video: "Race and Ethnicity"

1. Identify the problems in defining and using the term "race." Differentiate between race and ethnicity.

2. Identify examples of prejudice and discrimination that are shown in the video program. Explain whether or not it is easier to change prejudiced attitudes or discriminatory behaviors.

3. Identify some manifestations and examples of racism and discrimination against African-Americans.

4. Describe the illustrations presented in the video program of overt racism and discrimination against Asians in America.

5. Explain the human toll caused by discrimination as it appears in statistical patterns of educational attainment and professional employment and income. Be able to recognize examples.

RELATED ACTIVITIES

These activities may be used by your instructor as written assignments or as discussion topics. They may also be included as essay questions on your tests.

1. Are you a member of a minority group? Citing the elements of the definition given in the textbook as your reference, explain why you are or are not.

2. Explain the difference between genocide and expulsion. Cite at least one example of each to make your point.

3. What is your racial, ethnic, or national heritage? Would you say you have been assimilated or not? Explain, giving examples or evidence of your assimilation or non-assimilation.

4. Examine your own prejudices. Describe an incident in which you internally pre-judged someone who was of a racial, ethnic, religious, or national group different form your own. What were your judgments and feelings? Do not justify your judgments.

5. Interview a Latino person about his or her culture. Ask questions that will give you insight into that culture: questions about customs, values, family, beliefs, mutual support, art, and other special things about the culture. One of the aims of your interview is to discover elements of that culture upon which society could capitalize. Describe these beneficial aspects of the culture, and explain how you think our society as a whole could benefit.

6. If you have experienced prejudice or discrimination yourself, answer the questions that follow. If you have not, interview someone who has. Write out the responses to the following:
   A. Describe an instance in which you experienced prejudice or discrimination.
   B. What was said and done that showed the prejudice or discrimination?
   C. How did this make you feel?
   D. How did you react visibly?

PRACTICE TEST

The following items will help you evaluate your understanding of this lesson. Use the answer key at the end of the lesson to check your answers or to locate material related to each question.

Multiple-Choice

Select the one choice that best answers the question.

1. An ideology based on the belief that an observable, supposedly inherited trait, such as skin color, is a mark of inferiority that justifies discriminatory treatment of people is known as
   A. sexism.
   B. geneticism.
   C. racism.
   D. jingoism.

2. People who are singled out for differential and unequal treatment and who therefore regard themselves as objects of collective discrimination are known as
   A. ethnic groups.
   B. minority groups.
   C. racial groups.
   D. cultural groups.

3. The slaughter of thousands of tribal warriors in Africa by Europeans competing for colonial dominance is an example of
   A. segregation.
   B. assimilation.
   C. genocide.
   D. expulsion.

4. The expelling of Native Americans from their tribal lands by the white settlers is an example of
   A. segregation.
   B. assimilation.
   C. expulsion.
   D. amalgamation.

5. The ownership of a population, defined by racial, ethnic, or political criteria, by another population that not only can buy and sell members of the population but also has complete control over their lives is known as
   A. assimilation.
   B. segregation.
   C. genocide.
   D. slavery.

6. What type of segregation results from laws or other norms that force one people to be separate from others?
   A. De jure
   B. De facto
   C. Ipso facto
   D. Ex cathedra

7. Which of the following best represents an example of assimilation in Latin American societies?
   A. Races are allowed to mix in public places.
   B. Generations of intermarriage have occurred among the races and discrimination is largely absent.
   C. Different racial and ethnic groups attend school together but do not meet in casual social settings.
   D. Cultural pluralism is encouraged and a conscious attempt is made to encourage ethnic diversity.

8. The great barrier to a fully developed pluralism is
   A. population size.
   B. racism.
   C. cultural differences.
   D. deviance.

9. African-American people are good dancers.

   The above statement is an example of
   A. discrimination.
   B. segregation.
   C. Anglo-conformity.
   D. stereotyping.

10. Prejudice is an attitude, while discrimination is
    A. an unfair action.
    B. a stereotype.
    C. a negative feeling.
    D. an idea.

11. It is necessary to undo the effects of past discrimination.

    The above statement is an example of an argument in favor of
    A. ethnic nationalism.
    B. cultural pluralism.
    C. affirmative action.
    D. higher income taxes.

12. An individual feels a "free-floating" hostility that may be taken out on a convenient target or scapegoat.

    The above statement describes which theory of prejudice?
    A. Projection
    B. Anomie
    C. Frustration-aggression
    D. Functionalist

13. Which theory of intergroup hostility explains rather intense interactions of group members over a period of time that may cause perceptions that non-group members are inferior?
    A. Functionalist
    B. Conflict
    C. Interactionist
    D. Frustration-aggression

14. That white colonialists had a perceived need to use blacks for their own purposes and were able to enslave and sell Africans represents which theory of racial and ethnic inequality?
    A. Functionalist
    B. Interactionist
    C. Frustration-aggression
    D. Anglo-conformity

15. Which of the following does NOT fit Karl Marx's theory about what American workers would have to do to forge class loyalties?
    A. They would have to overcome divisions created by racial and ethnic differences.
    B. They would have to gain access to more complex technologies as a group and then use them.
    C. They would have to understand that they were being manipulated by the owners.
    D. They would have to be aware of how the owners were using strike-breakers of different racial and ethnic groups.

16. Which of the following is NOT a reason for the problem of lagging black mobility that has been identified by Stanley Lieberson and William Julius Wilson?
    A. Decrease in manufacturing jobs and increase in service-sector jobs is having a negative effect on the fortunes of African-American males.
    B. Recent research has shown that problems of African-American families today result chiefly from dissolution of black families by slavery.
    C. African-Americans experience more prejudice and discrimination, partly because they are more easily identified by their physical characteristics.
    D. High unemployment, coupled with the drug epidemic and other negative trends, contributes to the difficulty of young African-American couples in forming lasting relationships.

17. Which of the following is NOT an example of the negative emotions that are often part of prejudice?
    A. Mr. Donelee won't rent his houses to blacks; he thinks they're not a good risk.
    B. John is afraid when he sees a Hispanic person in a parking lot where he parks his car.
    C. Sam hates Asians; his father used to put them down unmercifully because he both feared and envied what he perceived as their business skill.
    D. Mrs. Edwards feels pity for her fourth-grade minority students and she feels superior to them.

18. Making judgments based on race is problematic because
    A. variation within and between groups is great.
    B. social class distinctions eliminate racial differences.
    C. distinctions between different races can be accurately made only by scientists.
    D. knowledge of cultural factors involved in racial differences is essential.

19. Which of the following is it easier to change?
    A. Attitudes
    B. Prejudices
    C. Behaviors
    D. Dispositions

20. Which of the following provides an example of racism and discrimination against African-Americans?
    A. Pacifying
    B. Feeling jealousy
    C. Rationalizing
    D. Redlining

21. Which of the following provides an example of overt racism against Asians in America?
    A. Stereotype of being lazy and shiftless
    B. Graffiti stating, "Jap go home"
    C. Belief that they remain drunk most of the time
    D. Perceptions of and resultant fear of their quick tempers

22. Which of the following most accurately portrays the effects of discrimination?
    A. About thirty percent of white men and only seventeen percent of black men hold professional and managerial positions.
    B. Suburban schools hire about the same percentage of minority teachers as represented in the general population.
    C. Many minority-owned businesses prefer to hire minorities.
    D. Many colleges and universities have the same expectations for minority students as for students of the majority group.

ANSWER KEY

The following provides the answers and references for the practice test questions. Objectives are referenced using the following abbreviations: T = text, R = Telecourse Guide Reading, and V = Video.

Answers	Lesson Goals	Objectives	References
1. C	1	T1	Kornblum, p. 399
2. B	1	T2	Kornblum, p. 403
3. C	2	T3	Kornblum, pp. 406-407
4. C	2	T4	Kornblum p. 407
5. D	2	T5	Kornblum, p. 407
6. A	2	T6	Kornblum, p. 409
7. B	2	T7	Kornblum, p. 411
8. B	2	T8	Kornblum, p. 415
9. D	3	T9	Kornblum, p. 416
10. A	3	T10	Kornblum, p. 416
11. C	3	T11	Kornblum, p. 420
12. C	4	T12	Kornblum, p. 421
13. C	4	T13	Kornblum, p. 422
14. A	4	T14	Kornblum, p. 422
15. B	4	T15	Kornblum, p. 423
16. B	5	T16	Kornblum, p. 425
17. A	3	R1	TG Reading
18. A	6	V1	Video
19. C	6	V2	Video
20. D	6	V3	Video
21. B	6	V4	Video
22. A	6	V5	Video

Notes and Assignments:

Lesson 16

Sex and Gender

LESSON ASSIGNMENT

Review the following assignment in order to schedule your time appropriately. Pay careful attention; the titles and numbers of the textbook chapter, the telecourse guide, and the video program may be different from one another.

Text:

Kornblum, *Sociology in a Changing World*,
Chapter 14, "Inequalities of Gender," pp. 436-467.

Reading:

There is no Reading for this lesson.

Video:

"Sex and Gender,"
from the series, *The Sociological Imagination*.

OVERVIEW

The new father, with several friends gathered around him, looks adoringly at his baby. The card in the slot at the foot of the crib is pink: It's a girl. From now until she dies, this particular human being will be touched, addressed, thought of, and otherwise treated quite differently than the newborn next to her, whose crib card is blue.

Right now, it is barely possible to distinguish the two diapered infants from each other. Yet this father's visions of playing with his child are based solely on her gender: She is a girl, so his visions don't include football games or wrestling. Instead, he sees himself watching her dress her doll and serving her tea with a tiny teapot hardly large enough for him to hold.

This scene is repeated hundreds of times a day in hospitals across the United States. Gender-role socialization begins at birth. There are behaviors many of us might consider almost biologically ordained—or prohibited—because of the sex of the individual. For example, the father wouldn't think of wrestling with this little girl, because it might hurt her delicate frame. Yet the Mundugumor, Tschambuli, and Arapesh peoples of New Guinea have three quite *dissimilar* definitions of the respective roles of men and women.

In this lesson you see how the socialization process unfolds quite differently for each gender. You also explore both the disparate, and the unfair, treatment of women in interpersonal relationships and in the opportunity system of our society.

In the video program you meet some women who describe aspects of their biographies: how gender made a difference in their lives. Their stories illustrate the results of the differential treatment of the sexes at a personal level. In addition, you explore the broader social implications of unequal treatment and the ways women historically have responded to it.

Why are women and men so different? What kinds of obstacles of full participation in our society have women faced? Why? What does sociology have to say about gender and about inequality based on gender? These questions launch us into this fascinating and important study of the importance of gender—a study that should further stimulate your sociological imagination.

LESSON GOALS

Upon completing this lesson, you should be able to:

1. Identify the major changes in sexuality over the past 300 years.
2. Analyze how gender and gender stratification affect the course of people's lives.
3. Analyze gender roles and gender stratification, and how these have been manifested.
4. Analyze the social changes that have occurred as a result of the women's movement.

5. Identify social changes that have occurred in the industrial nations of North America, Europe, and Oceania with the increasing numbers of women in the work force.

6. Analyze how members of society historically have treated each other differently based on gender and sexual differentiation; analyze the personal and social results of that treatment.

TEXTBOOK OBJECTIVES

The following textbook objectives are designed to help you get the most from the text. Review them, then read the assignment. You may want to write notes to reinforce what you have learned.

Text: Kornblum, *Sociology in a Changing World*, Chapter 14.

1. Identify the range of sexual behaviors and norms throughout the world. Be able to define: polygny, incest taboo; marriage; heterosexuality; homosexuality; bisexuality; transsexuality; sex; hermaphrodites; and primary and secondary sexual characteristics.

2. Define gender stratification, and explain how it is reflected in women's participation in professions.

3. Describe how norms regarding gender are linked to the life course established by a society, specifically in eighteenth-century societies.

4. Define gender roles. Give examples.

5. Describe the scientific evidence about innate biological and psychological bases for sex roles. Explain what Margaret Mead's research indicated about the behavior of males and females.

6. Describe how people are socialized to gender roles over time in industrial societies. Analyze how this affects interaction patterns of boys and girls in the United States.

7. Analyze how social structure affects the inequalities and differences between men and women in industrial societies. Be able to recognize examples of the impact of social structure on these inequalities and differences.

8. Define sexism, and identify some historical manifestations of it. Be able to recognize examples.

9. Describe the women's movement and how it has affected social change.

10. Identify the impact World War II had on women in the workplace and changes occurring after the war.

11. Identify problems encountered by women in the labor force. Define "pink-collar ghettos" and "the second shift."

VIDEO OBJECTIVES

The following video objectives are designed to help you get the most from the video segment of this lesson. Review them, then watch the video. You may want to write notes to reinforce what you have learned.

Video: "Sex and Gender"

1. Differentiate between sex and gender. Differentiate between patriarchal, non-patriarchal, and matriarchal societies.

2. Virginia Woods is the silver-haired woman interviewed on the video program. Describe and exemplify the differences in roles and career opportunities of the women in Virginia Woods' generation and those who lived in her mother's generation. Identify the technological changes that occurred during Virginia Woods' time.

3. Explain why there is inequality between men and women today and why women today are in crisis.

4. Define the "second shift." Identify some ways the problems related to the "second shift" could be dealt with.

RELATED ACTIVITIES

These activities may be used by your instructor as written assignments or as discussion topics. They may also be included as essay questions on your tests.

1. Record the poems, sayings, phrases, and words used to describe little boys and little girls that you have heard from parents, relatives, friends, and teachers throughout your life. List at least three cultural norms about the genders indicated by what you have written.

2. List, explain, and give three examples of ways the United States' culture reinforces violence toward women. (Example: Mass media continues to televise programming depicting women as weak; an example is Edith in *All in the Family* or Peg in *Married with Children*.)

3. Describe some signs that American society is a patriarchal society.

PRACTICE TEST

The following items will help you evaluate your understanding of this lesson. Use the answer key at the end of the lesson to check your answers or to locate material related to each question.

Multiple-Choice

Select the one choice that best answers the question.

1. Which of the following statements regarding the "cause" of homosexuality is true?
   A. There is insufficient evidence to resolve this issue.
   B. Homosexuality is a "lifestyle choice."
   C. Homosexuality has genetic origins.
   D. Homosexuality is a combination of genetic patterning and cultural choices.

2. Which of the following best illustrates the way gender stratification is reflected in women's participation in the professions?
   A. By 1956 women had reached numerical parity with men in the professions.
   B. By 1986 the weekly earnings of professional women were more or less equal to earnings of men.
   C. Women's incomes have not matched men's; women's participation in the professions is more or less equal to that of men.
   D. Men and women are not yet equal, at least in terms of earnings, prestige, and power.

3. The origins of gender inequality in most modern societies can be traced to their
   A. industrial periods.
   B. pre-industrial periods.
   C. feudal periods.
   D. agrarian periods.

4. Sets of behaviors that are considered appropriate for individuals of a particular gender are called gender
   A. ratios.
   B. roles.
   C. values.
   D. handicaps.

5. The social sciences
   A. tend to support the biological explanations for different roles and temperaments of men and women.
   B. call into question the psychological explanations for different roles of men and women.
   C. support the notion that men's and women's abilities and personalities are different.
   D. emphasize the psychological explanations of role differences in women and men.

6. One of the consequences of girls and boys learning different roles, for example that boys spend more time playing team sports than girls, is that women
   A. are more likely to be socialized into the "feminine" roles of mother, teacher, and secretary.
   B. because of their superior relational skills have greater opportunity in the skilled labor market than do men.
   C. and men are more suited to participation in heterogeneous society than their peers in previous times.
   D. tend to better fit the more feminine demands of the contemporary work force.

7. Which of the following best illustrates the impact of social structure - as opposed to socialization - on the inequalities and differences between the roles of men and women?
   A. Programming girls that they should be good at domestic roles
   B. Segregating games for boys and girls
   C. Implementing school rules that permit coeducational sports
   D. Teaching cultural norms about feminine and masculine behaviors

8. The objectification of women in the role of the beauty contestant is an example of
   A. sexism.
   B. sex ratio.
   C. sexual identity.
   D. gender validity.

9. Which of the following is NOT true of the women's movement in the late nineteenth and early twentieth centuries?
   A. Women whose status and education were improving combined a number of existing organizations, forming a movement.
   B. Women organized to gain full citizenship rights.
   C. Women demanded greater control over reproduction through family planning and birth control.
   D. Women who participated were known as suffragettes.

10. An achievement of the women's movement of the 1970s resulted in greater enforcement of federal laws
    A. enacting complete gender equality.
    B. enforcing the right of women to run for public office.
    C. making birth control available in public clinics.
    D. barring gender-based discrimination in employment.

11. Which of the following is NOT true about World War II's effect on women's labor force participation?
    A. World War II marked a significant turning point in women's labor force participation.
    B. All employed women were "bumped" from jobs in factories and offices.
    C. Female employment was increasing at a far faster rate than male employment.
    D. War widows and single women had to continue earning wages.

12. Which of the following occupations does NOT exemplify "pink-collar ghettos"?
    A. Secretarial work
    B. Physician
    C. Child care
    D. Dental assistant

13. In which type of society do women share equally with men in positions of leadership and authority?
    A. Patriarchal
    B. Non-patriarchal
    C. Matriarchal
    D. Paternal

14. Virginia Woods (1914-) was the silver-haired woman interviewed on the video program on sex and gender.

    Which of the following best describes the kinds of opportunities possessed by women of Virginia Woods' generation?
    A. They were entirely trapped at home on the farm.
    B. They were able to pursue careers in a limited number of areas, such as teaching.
    C. The industrial revolution opened up many careers in industrial plants that women were called on to fill.
    D. Women could basically do anything they wanted to with their lives.

15. Women experienced inequality initially due to
    A. crowding into a few professions, thus lowering the pay rate of these professionals.
    B. supervisory personnel without managerial techniques to maximize women's potential.
    C. complex bureaucracies unable to accommodate the talents of women.
    D. failure to identify the true basis of discrimination in the workplace.

16. What does the video program on sex and gender describe as a way to cope with the problems related to the "second shift"?
    A. Change management styles of those who are in authority in the workplace.
    B. Adjust the role of men so that they do not distance themselves from the work at home.
    C. Force men to speak more respectfully to women in all social situations.
    D. Make cultural images of women more suited to modern life.

ANSWER KEY

The following provides the answers and references for the practice test questions. Objectives are referenced using the following abbreviations: T = text, R = Telecourse Guide Reading, and V = Video.

Answers	Lesson Goals	Objectives	References
1. A	1	T1	Kornblum, pp. 446-447
2. D	2	T2	Kornblum, p. 448
3. C	2	T3	Kornblum, p. 449
4. B	3	T4	Kornblum, p. 447
5. B	3	T5	Kornblum, p. 448
6. A	3	T6	Kornblum, p. 451
7. B	3	T7	Kornblum, p. 451
8. A	3	T8	Kornblum, pp. 452-453
9. A	4	T9	Kornblum, p. 454
10. D	4	T9	Kornblum, p. 454
11. B	5	T10	Kornblum, p. 457
12. B	5	T11	Kornblum, p. 459
13. B	6	V1	Video
14. B	6	V2	Video
15. A	6	V3	Video
16. B	6	V4	Video

Notes and Assignments:

Lesson 17

Aging

LESSON ASSIGNMENT

Review the following assignment in order to schedule your time appropriately. Pay careful attention; the titles and numbers of the textbook chapter, the telecourse guide, and the video program may be different from one another.

Text:

Kornblum, Sociology in a Changing World,
Chapter 15, "Inequalities of Youth and Age," pp. 468-495.

Reading:

There is no Reading for this lesson.

Video:

"Aging,"
From the series, *The Sociological Imagination*.

OVERVIEW

The man sits in his big stuffed easy chair, which is perched on a six-inch pedestal to ease the arthritic pain he feels each time he gets up. He proudly wears his bright orange hat with these words emblazoned on the front: "Don't ask me to do a damn thing, I'm retired!" Tyrone sold his car four years ago and gets out of the house only to go to his doctors.

In a nearby suburb, another man drives with his wife to the mall, where they will make their daily jaunt. Larry and his spouse of fifty years will see and speak to several friends during their walk. Later that day they will go help at the church's food bank and attend the AARP meeting in the evening.

Not far away from Tyrone and Larry, a third man is dressed and ready to leave for work. He probably will work only a few hours, then come home and take a nap. But Paul needs to work to feel valued and useful—whether or not he needs the money to live on.

Tyrone, Larry, and Paul are all in their late seventies, and all three suffer from various chronic health problems. Their modes of adaptation, however, are quite different:

1. Tyrone has disengaged from almost all activity, going through what Robert Atchley calls a "consolidation" of commitments and a redistribution of available energy as a way of coping with lost roles, activities, or capacities.

2. Larry and his spouse, on the other hand, have adapted by "substitution"; that is, they have—as much as possible—replaced pre-retirement activities with other (post-retirement) activities.

3. Paul's method of adaptation is to continue to do what he has always done—but at a slower pace.

In spite of stereotypes and inaccurate beliefs among many people, there is no single model of adaptation, or retirement, or of otherwise dealing with the aging process. Some popular periodicals tend to present a picture of the aging as a powerful, wealthy group. There is some truth to this depiction, since it is a myth that all old people are helpless and poor. However, much ageism and age discrimination still exist, and many older citizens are isolated, lonely, poverty stricken, and homeless.

The video program shows you a variety of older people and the differing ways they have adapted to the aging process. It depicts some of the popular stereotypes and provides a sociological interpretation of age, age stratification, and ageism.

As you and those close to you grow older, it is more important than ever to develop a sociological imagination about aging. This lesson helps you to better understand and cope with the social realities of growing old in America.

LESSON GOALS

Upon completing this lesson, you should be able to:

1. Analyze the process of aging and the life course.
2. Analyze age stratification and ageism.
3. Analyze the sociological perspectives on age stratification and the aging process.
4. Analyze how society treats older people.

TEXTBOOK OBJECTIVES

The following textbook objectives are designed to help you get the most from the text. Review them, then read the assignment. You may want to write notes to reinforce what you have learned.

Text: Kornblum, *Sociology in a Changing World*, Chapter 15.

1. Define age stratification and age strata. Be able to identify indicators of age strata.
2. Define life course and rites of passage, and give examples of the latter.
3. Explain how the age cohorts of the baby boom and the baby bust illustrate a population pyramid. Describe the demands on a society's institutions as these cohorts mature.
4. Define life expectancy, and describe how it is related to gender.
5. Describe disengagement theory and activity theory. Be able to recognize examples.
6. Define and give examples of ageism. Describe Robert Butler's observations about ageism.

7. Describe the mismatch between the strengths and capacities of older people and their roles in society according to well-known gerontologists Matilda White Riley and John W. Riley, Jr.

8. Describe the hospice movement. Explain its purpose and contributions.

VIDEO OBJECTIVES

The following video objectives are designed to help you get the most from the video segment of this lesson. Review them, then watch the video. You may want to write notes to reinforce what you have learned.

Video: "Aging"

1. Identify the stereotypes about older people. Explain the root of these stereotypes.

2. Explain how discrimination against older people is connected with stereotypes and negative attitudes toward aging. Describe and exemplify several ways that older people experience discrimination.

3. Identify some of the difficulties and positive aspects involved in growing old. List some compensations for growing old, and describe how older people can find meaning in their lives and make aging a positive experience.

RELATED ACTIVITIES

These activities may be used by your instructor as written assignments or as discussion topics. They may also be included as essay questions on your tests.

1. Describe a rite of passage you have gone through. What were your feelings about yourself before and after? How did people treat you differently after the experience?

2. Record comments you hear during the next week or two about older people or the process of aging. What sorts of images of the aging process emerge from these comments? Give examples and summarize your findings.

3. Stereotypes are overgeneralized images of people in a particular category. Even when there may be a grain of truth in a stereotype, it inaccurately portrays all the people it is supposed to represent.

   Describe a common stereotype of older people. What is the basis, the "grain of truth," in that stereotype? What evidence proves the error of that stereotype?

4. List three stereotypes you hear about older people. Explain the basis for each stereotype.

PRACTICE TEST

The following items will help you evaluate your understanding of this lesson. Use the answer key at the end of the lesson to check your answers or to locate material related to each question.

Multiple-Choice

Select the one choice that best answers the question.

1. Which of the following is NOT an indicator of age strata?
   A. As people grow older they may experience problems with eyesight due to the hardening of the retina.
   B. People in different age strata command different amounts of scarce resources, such as wealth, power, and prestige.
   C. Numerous laws establish inequalities between youth and adults, such as the right to vote or incur debt.
   D. Hundreds of thousands of children and teenagers do not receive care to offset their unequal status under the law.

2. Graduation is an example of
   A. ageism.
   B. age domination.
   C. disengagement.
   D. a rite of passage.

3. Rapid increases in the birthrate from about 1945 through the early 1960s resulted in the
   A. hippie generation.
   B. age boom.
   C. industrialization population pyramid.
   D. baby boom.

4. Life expectancy is the
   A. average number of years that a member of a given population can expect to live beyond his or her present age.
   B. amount of time between birth and old age that is considered the productive phase of one's life.
   C. difference between average length of life and death rate in a society.
   D. average of the birthrates of a population subtracted from the average of the death rates of that same population.

5. Upon retirement, the chief executive officer of a Fortune 500 firm suffers from a sense of loneliness and loss.

   The description is an example of age:
   A. disengagement theory.
   B. activity theory.
   C. role deprivation theory.
   D. retirement theory.

6. The ideology that justifies prejudice or discrimination based on age is known as
   A. ageism.
   B. gerontology.
   C. geriatric fundamentalism.
   D. age mythology.

7. As people live longer they
   A. move in with their children.
   B. find themselves living alone.
   C. move into retirement homes.
   D. live in exotic places.

8. In a hospice, the most relevant social unit is the
   A. hospital itself.
   B. medical staff, including sensitive and well-trained nurses.
   C. patient and his or her surrounding social group.
   D. insurance companies that provide for the care of the dying patient.

9. Which of the following is NOT a stereotype about older people?
   A. Older people are politically apathetic.
   B. Older people are inactive and conservative in both lifestyle and politics.
   C. Older people are sexually inactive.
   D. Older people are completely disengaged or retired.

10. Which of the following is an attitude toward aging that accompanies efforts to segregate older people?
    A. Younger people's fear of their own mortality
    B. Instrumentalism among the working populace
    C. Empathy for those having problems with growing older
    D. Humiliation among the young

11. Which of the following is NOT spoken of in the video program as a
    A. Loneliness accompanying loss of mate
    B. Subordinate position in the political system
    C. Failing health, such as problems with cataracts
    D. Pain of having to leave your home

ANSWER KEY

The following provides the answers and references for the practice test questions. Objectives are referenced using the following abbreviations: T = text, R = Telecourse Guide Reading, and V = Video.

Answers	Lesson Goals	Objectives	References
1. A	1	T1	Kornblum, p. 470
2. D	1	T2	Kornblum, p. 470
3. D	1	T3	Kornblum, p. 472
4. A	1	T4	Kornblum, p. 476
5. A	1	T5	Kornblum, p. 488
6. A	2	T6	Kornblum, p. 486
7. B	3	T7	Kornblum, p. 485
8. C	3	T8	Kornblum, p. 489
9. A	4	V1	Video
10. A	4	V2	Video
11. B	4	V3	Video

Notes and Assignments:

Lesson 18

Education

LESSON ASSIGNMENT

Review the following assignment in order to schedule your time appropriately. Pay careful attention; the titles and numbers of the textbook chapter, the telecourse guide, and the video program may be different from one another.

Text:

Kornblum, *Sociology in a Changing World*,
Chapter 18, "Education and Communications Media," pp. 568-589.

Reading:

There is no Reading for this lesson.

Video:

"Education,"
From the series, *The Sociological Imagination.*

OVERVIEW

The grimace on the father's face was no less strained than his son's. The teacher had said that the computer program assignment was simple and should be easy to set up. Yet their efforts to do the assignment met with nothing but frustration. Ed asked his son, Paul, why he hadn't gotten better instructions.

"Well, for one thing, the teacher was mad at me. She said I was impolite and interrupted too much. And besides, I couldn't get to a computer for the teacher to work with me. There are twenty-eight students and we only have four computers."

"Why only four? Do you interrupt like she said?"

"Our teacher told us she had begged the principal for more, but the school board had turned her down. They said they'd try to get a business to donate some to us. And, yes, I guess I do give her a hard time. School seems like a jail to me, Dad."

Ed chose to ignore the last statement. He could remember his own rebellion against the rigidity of school. Angry with the lack of resources of the small school district, Ed muttered, "How do they expect you to be able to find a job if you don't learn the latest technology? I'm going to the next school board meeting and give them a piece of my mind!"

He turned back to Paul, saying, "Where's your English assignment? I know I can help you with that. Grandpa made sure I knew the rules of grammar and he made me write every night. Why did you wait 'til you were almost failing two classes to talk to me about your troubles?" Ed's confusion was thorough.

If Paul's father follows through with his pledge to become more involved with the school, perhaps without realizing it, he will be joining a national movement to reform education. But he will soon discover that the solution to the problem of unequal resources for smaller and poorer school districts lies beyond any dictums of his local school board. He also will find out that forging new relationships between teachers and students, between businesses and schools, is extremely difficult.

The school system is a bureaucracy. Unequal funding of schools is a result of the social stratification system and the social and economic arrangements of his community and state. In the heat of his discussion with his son, Ed also forgot about the fact that his state officials were presently under a court order to equalize funding among rich and poor school districts. And the governor had just ordered the sixth special legislative session to try to solve the problem as a result of his "no new taxes" vetoes.

There also is more to the relationship between teacher and student than his son's frank admission of defiance. If Ed knew the sociological research on the subject, he would understand that Paul is not just an ornery kid but part of a peer group. And this "adolescent society" is itself a social response to the pressures placed upon teenagers by parents, school official, and broad economic conditions.

Who goes to the schools and colleges? Who controls them? What are their functions and goals, and who decides what the goals should be? Why are some schools failing and others succeeding? These are some of the questions and issues you encounter in this lesson as you develop your sociological imagination about the institution of education.

LESSON GOALS

Upon completing this lesson, you should be able to:

1. Analyze the nature and function of the institution of education, and evaluate the diverse goals schools have had over time.

2. Analyze educational attainment and achievement and the socioeconomic factors involved.

3. Analyze the structure of educational institutions and the qualities of the teacher-student relationship, and evaluate changes in American schools.

4. Analyze the relationship between education and other social institutions.

TEXTBOOK OBJECTIVES

The following textbook objectives are designed to help you get the most from the text. Review them, then read the assignment. You may want to write notes to reinforce what you have learned.

Text: Kornblum, *Sociology in a Changing World*, Chapter 18.

1. Explain the interactionist and conflict views of education, and give examples of research from each.

2. Describe the conclusions of James Coleman and other sociologists about "adolescent society" and youth culture. Identify the costs or negative aspects of these phenomena for adolescents and for society.

3. One of the goals of education in America is educating for citizenship. Describe these goals, how they are contradictory, and why they are almost impossible to achieve.

4. Define educational attainment. Describe historical patterns of educational attainment among different segments of the population and the problems associated with them.

5. Describe recent research on school achievement, comparing American and Asian schoolchildren. Explain the practical implications for social change in American schools.

6. Describe the findings of Lourdes Diaz Soto, Sandra Baum, and Ray C. Rist with regard to educational achievement and social mobility.

7. Describe the shape and importance of the teacher-student relationship in American schools. Explain what research has revealed about student response to conventional classrooms.

8. Identify what makes change difficult for schools and teachers. Include the difficulty of desegregation.

VIDEO OBJECTIVES

The following video objectives are designed to help you get the most from the video segment of this lesson. Review them, then watch the video. You may want to write notes to reinforce what you have learned.

Video: "Education"

1. Describe and exemplify the pressures placed on education by the social institutions of religion, science, and the family. How is education vulnerable to political pressure? Explain how some school districts are placed in an advantageous position because of funding sources and procedures.

2. Explain how successful business has been in predicting future labor markets, and be able to recognize examples. What are the interests of business as it seeks to influence education? Describe the issues and problems surrounding control of schools by business.

3. Describe the problems connected with running a school like a business. Discuss the professor's conclusions in the video program about the value of business pressure and involvement in schools.

RELATED ACTIVITIES

These activities may be used by your instructor as written assignments or as discussion topics. They may also be included as essay questions on your tests.

1. Thinking back on your high-school days, were you or your friends part of an "adolescent society," as described by James Coleman and other sociologists? Did you experience a "compulsive conformity" and loyalty to a peer group? If you cannot identify with these questions for yourself, do you see these phenomena happening with your children or other teenagers?

2. List and explain three positive and three negative factors you experienced in your education (school).

3. Do you agree with the author of the textbook that the expectations that education can solve social problems seem impossible to meet? Explain why or why not.

4. The research findings of Ray C. Rist go contrary to the popular belief that education is the great leveler and the solution to social problems associated with social stratification and discrimination. Ask at least five people who are college students and five who are not about their beliefs about education's ability to provide social mobility to all segments of society. Summarize your findings and describe how they do or do not support Jenck's research conclusions.

PRACTICE TEST

The following items will help you evaluate your understanding of this lesson. Use the answer key at the end of the lesson to check your answers or to locate material related to each question.

Multiple-Choice

Select the one choice that best answers the question.

1. The interactionist viewpoint sees schools as
   A. a specialized structure with a special function.
   B. an instrument to achieve equality.
   C. a set of behaviors.
   D. an embodiment of the ideals and values of the society.

2. In his work *The Adolescent Society*, James Coleman found that
   A. schools create a social world for adolescents that is separate from adult society.
   B. complexity of industrial societies is forcing adolescents to depend more heavily on family relationships.
   C. schools are providing the support and individual attention increasingly needed by adolescents.
   D. promiscuity among adolescents has increased dramatically as a result of the pressures of living in an industrial society.

3. Which of the following is true about the goals of education for citizenship within American society?
   A. Schools have been expected to teach values but not to shape society itself.
   B. Schools have been expected to create equal opportunity and prepare new generations of citizens to function in society.
   C. That education should help children maintain their cultural identity is almost universally accepted.
   D. Nearly all school districts throughout the country have adopted the goal of encouraging critical thinking skills in their students.

4. National service - the full-time undertaking of public duties by young people - is
   A. a likely legislative priority in the near future.
   B. a goal held by reformers who believe that formal education should include two years of college.
   C. an aspect of the back-to-basics movement in the United States.
   D. an attempt to link formal schooling with programs designed to address social problems.

5. Which of the following best describes historical patterns of educational attainment among different segments of the population in America?
   A. Average Americans today actually have less education than average Americans of the early 1940s.
   B. Whites are more likely than blacks to complete high school.
   C. Fewer women than men complete high school.
   D. Due to affirmative action, blacks are more likely than whites to complete four years of college.

6. Recent research reveals that the mathematics scores of American first-graders were
   A. higher than those of Asian first-graders.
   B. lower than those of Asian first-graders.
   C. the same as those of Asian first-graders.
   D. higher than Asian second-graders.

7. What has Ray C. Rist concluded about the system of public education in the United States?
   A. Teachers' expectations had little effect on student performance when compared to their ability to teach.
   B. Education increases social mobility.
   C. Public education is designed to aid in the establishment of a more equal society.
   D. Public education is designed to aid in the perpetuation of the social and economic inequalities found in the United States.

8. According to research done on the teacher-student relationship in American schools, it is evident that
   A. teacher-student relationships are not as important as the investment in educational technology.
   B. good teachers can make a big difference in children's lives.
   C. more investment in the quality of the teaching is needed in secondary schools than in primary schools.
   D. conventional classrooms provide sufficient opportunity for the student to become actively involved in the learning process.

9. What is it about school systems that makes them highly resistant to change?
   A. Old facilities that are difficult to adapt to new methods
   B. Uncreative teachers
   C. Bureaucratic organization of school systems
   D. Fragmented nature of the curricula

10. Which of the following best illustrates the fact that some schools are in advantageous positions because of funding sources and procedures?
    A. High real estate values can provide thirty times as much income as low real estate values.
    B. Poor school districts compensate by hiring more student-oriented teachers.
    C. Education funded by sales taxes is more efficient than education funded by property taxes.
    D. Larger school districts are less able than smaller school districts to attract well-qualified professional educators.

11. One of the main interests of business groups that seek to influence education is to
    A. provide a sound tax base for educational institutions.
    B. stimulate young people to broaden their intellectual horizons.
    C. make the United States more competitive in world markets.
    D. flood the labor market, thereby reducing wages for workers.

12. While businesses measure success and productivity by their profits, schools that are run like businesses find it difficult to
    A. hire administrators with certification.
    B. measure good teaching objectively.
    C. deal with financial matters.
    D. pay the salaries of bureaucrats.

ANSWER KEY

The following provides the answers and references for the practice test questions. Objectives are referenced using the following abbreviations: T = text, R = Telecourse Guide Reading, and V = Video.

Answers	Lesson Goals	Objectives	References
1. C	1	T1	Kornblum, p. 570
2. A	1	T2	Kornblum, p. 573
3. B	1	T3	Kornblum, p. 573
4. D	1	T3	Kornblum, p. 574
5. B	2	T4	Kornblum, pp. 574-575
6. B	2	T5	Kornblum, p. 578
7. D	2	T6	Kornblum, p. 582
8. B	3	T7	Kornblum, p. 586
9. C	3	T8	Kornblum, p. 586
10. A	4	V1	Video
11. D	4	V2	Video
12. B	4	V3	Video

Notes and Assignments:

Lesson 19

Family

LESSON ASSIGNMENT

Review the following assignment in order to schedule your time appropriately. Pay careful attention; the titles and numbers of the textbook chapter, the telecourse guide, and the video program may be different from one another.

Text:

Kornblum, *Sociology in a Changing World,*
Chapter 16, "The Family," pp. 498-531.

Reading:

There is no Reading for this lesson.

Video:

"Family,"
from the series, *The Sociological Imagination.*

OVERVIEW

"If you don't stop kicking me, I'll tell my Uncle Elbert," shouts Tim.

"Oh, he's not your uncle, and anyway, my daddy is bigger," Tanya responds heatedly.

Tim retreats into his house and runs sobbing to his mother, "Isn't Uncle Elbert my real uncle?"

"Well, he's not blood, but he's family," his mother answers. She explains that Elbert is Dad's old high school buddy and that "Uncle" Elbert, as Tim has been taught to call him, is just staying with Tim's family for a year or two while going to college to be retrained after being fired from the auto plant.

Tim is puzzled about the "blood" part of it, but he is comforted. What is important to Tim is that "Uncle" Elbert is family.

But is he? What is a family, or what constitutes "family"? In fact, what is kinship? Tim had laughed at the little girl dancing and parading before her father and mother on the television sitcom the other night. He had thought to himself then, "Is this the way families are supposed to be: two parents and their children living along in a house?" But now he wondered, aren't the three elderly sisters living in the house across the street also a family? And what about Tanya's family, which is only her brother, her widowed mother, and Tanya herself? Even more baffling are Jim and Alvin, the gay couple who live two doors down. Or the Indian couple down the street, whose household includes her mother, her two brothers, and a distance cousin—living here for more than two years now. Are all these families? If not, what would we call them?

The next question to ponder, for Tim and for us: Is it kinship or is it interaction that makes a group a family? Certainly there are brothers, sisters, and parents who refuse to speak to each other because of intense hurt, outright abuse, or differences in lifestyles or values. Are they still a family?

On the other hand, people who are unrelated by blood ties but live near each other may continually rely on each other for daily emotional and physical support. Unlike the brothers and sisters who do not communicate, they interact frequently. They may consider themselves family. But are they?

Families are the foundation of society. They provide the basic unit of social organization that makes it possible for a society to be a social system. The larger question we need to consider, then, is this: How do families vary in form and function across cultures and down through time?

After studying this lesson, the answers to the questions raised here will become clearer. In the video program you encounter several families, each a different type, each approaching life and its problems in its own way. And you see one family trying to cope with special difficulties and dysfunction.

Before you begin this lesson, you probably think it a simple task to define a "family." In fact, you no doubt know a lot about families from your own experiences. Although this knowledge will help you, you also are challenged here to think about families in new ways—and to develop your sociological imagination on this important, yet continually perplexing, subject.

LESSON GOALS

Upon completing this lesson, you should be able to:

1. Analyze the different types of families and examine how families have changed in form and function over time.
2. Analyze the process of mate selection, the components of love, and the sources of marital instability.
3. Analyze the basic sociological perspectives on the family
4. Analyze how the families on the video program provide for emotional support and individual growth, as well as how they cope with difficulties.

TEXTBOOK OBJECTIVES

The following textbook objectives are designed to help you get the most from the text. Review them, then read the assignment. You may want to write notes to reinforce what you have learned.

Text: Kornblum, *Sociology in a Changing World*, Chapter 16.

1. Identify the essential needs any society must meet. Explain how institutions—including the family—in simple and complex societies fill needs in different ways.
2. Discuss the traditional definitions of family and kinship. Explain why these definitions are not completely adequate.
3. Differentiate among consanguineous attachments, conjugal relations, nuclear family, family of orientation, and family of procreation. Be able to recognize examples of each.
4. Describe what has happened to the traditional household consisting of two parents and their children.

5. Identify the stages of the "family life cycle" according to Paul Glick and discuss how "age at first marriage" has changed.
6. Identify the parental issues faced by both biological and "step" parents.
7. Explain how marriage is an exchange. Identify factors that are considered in arranging a marriage.
8. Differentiate between endogamy, exogamy, and homogamy.
9. Describe and exemplify the theories of complementary needs and emotional reciprocity.
10. Describe the social-scientific evidence about the early years of family formation and about trial marriages.
11. Describe Robert Sternberg's research on the components of love.
12. Identify the sources of marital instability.
13. Describe the sociological evidence about the impact of divorce on adults and children.
14. Identify the basic contradiction and the core problem of many American families. Describe the "devitalized" couple and the "empty shell" marriage.
15. Explain the functionalist view of changes in families. Describe how changes in American values have affected families.
16. Describe the overall poverty rate for African-American families, compared to that of other races. Explain sociologists' conclusions about the reasons for the rapid rise in black female-headed households.

VIDEO OBJECTIVES

The following video objectives are designed to help you get the most from the video segment of this lesson. Review them, then watch the video. You may want to write notes to reinforce what you have learned.

Video: "Family"

1. Identify what makes families vital and desirable to human kind. Explain whether or not families remain the same throughout history.

2. Compare and contrast the ways the families portrayed in the video program have influenced and changed their individual family members and how they give emotional support. Describe how being a single parent has changed Carla Cargile.

3. Describe the dysfunctions, strains, and difficulties mentioned by Patrick and Hillary Shaw. Explain how they are coping with them and describe what a family can do to become more functional.

4. Identify what historical changes have taken place in the family as a unit. Discuss the importance and functions of families today and how the family will change in the future.

RELATED ACTIVITIES

These activities may be used by your instructor as written assignments or as discussion topics. They may also be included as essay questions on your tests.

1. If your family is a "blended" family involving step-parents, step-children, or "half" brothers and/or sisters, write about the most difficult situations/problems you have encountered in your "new" family. Your perspective may be that as a parent or child—use whichever role you fulfill. How effectively did you deal with each situation? What has been the most difficult to resolve? What lessons have you learned from being in a "blended" family?

2. Briefly describe at least one specific example of how your family of orientation or family of procreation has provided for each of the six essential needs listed in the textbook.

3. Describe two families you know that fit the following categories:
   A. a conjugal family, and
   B. a nuclear family that is not a conjugal family.

4. Have you seen pressures placed on yourself, others in your family, or someone you know to marry within a specific cultural (endogamy) or class (homogamy) group? Describe these pressures.

5. Examine a love relationship of your own or of a confidant. Describe the relationship in terms of the degree to which Sternberg's three components were present. (You need not be overly graphic, and you may fictionalize to "protect the innocent.")

6. Use people from a book, article, movie, or television program to describe a conflict-habituated couple and an "empty shell" marriage.

7. Describe the ways your family provided you with social placement in society.

8. Did you get sufficient emotional support from your family of orientation? How has that affected your relationships and your sense of self today?

9. How was the labor divided in your family of orientation? Explain how the pattern set there affects your family roles today or the roles you are likely to play in the future.

PRACTICE TEST

The following items will help you evaluate your understanding of this lesson. Use the answer key at the end of the lesson to check your answers or to locate material related to each question.

Multiple-Choice

Select the one choice that best answers the question.

1. Which of the following is NOT an essential need all societies must meet?
   A. Distribution of goods and services
   B. Provision of emotional support
   C. Replacement of members
   D. Control of members

2. The role relations among people who consider themselves to be related by blood, marriage, or adoption are referred to as
   A. role expectations.
   B. kinship.
   C. blood ties.
   D. group relations.

3. Two or more people related by consanguineous or conjugal ties or by adoption who share a household are commonly referred to as
   A. a conjugal family.
   B. a nuclear family.
   C. an extended family.
   D. a fictive family.

4. The traditional household consisting of two parents and their children is now becoming
   A. a thing of the past.
   B. a less common occurrence.
   C. a conjugal family.
   D. an extended family.

5. In Paul Glick's "family life cycle," the "family dissolution" stage refers to
   A. family formation.
   B. remarriage of partners.
   C. divorce.
   D. death of a spouse.

6. When stepfamilies are formed
   A. family members remain in former roles.
   B. you find the family culture developed from the man's culture.
   C. the task of working out a mutually acceptable concept of family is usually stressful.
   D. adults and children find a smooth transition into the "new" family.

7. In India transactions involving the arrangement of marriages are based on
   A. the degree of intimacy experienced by the couple about to be married.
   B. family prestige, wealth, and other social and personal factors.
   C. romantic love.
   D. the length of time the future bride and groom have known one another intimately.

8. Marriage outside the group is termed
   A. endogamy.
   B. homogamy.
   C. polygamy.
   D. exogamy.

9. The theory that relationships flourish when people feel satisfied is known as the theory of
   A. romantic love.
   B. emotional reciprocity.
   C. complementary needs.
   D. complementarity.

10. Most difficult for a couple are the years immediately following the
   A. birth of a second child.
   B. formalization of a marriage relationship.
   C. realization that one partner is intellectually inferior to the other.
   D. retirement of the male partner.

11. According to Robert Sternberg, the shared sense that a couple can reveal their innermost feelings to each other even as those feelings change is
   A. commitment.
   B. passion.
   C. intimacy.
   D. negotiation.

12. Which of the following is NOT a source of marital instability?
   A. Waiting too long to marry
   B. Marrying on the rebound
   C. Becoming parents after two years of marriage
   D. Depending on an extended family for income

13. Children of divorced parents tend to
   A. adjust better if their parents have been married only a short time.
   B. exhibit a variety of psychological problems.
   C. handle difficulties of the marital dissolution better if one parent had experienced a previous divorce.
   D. Excel in school in order to curry the favor of unhappy parents.

14. The contradiction that is inherent in the institution of the family is between the
    A. religious values underlying family life for centuries and the values found in society.
    B. aggressive husband and the passive wife.
    C. values taught to children by mass media and the norms promoted by school and religion.
    D. need to maintain individuality while providing love and support within a set of interdependent relationships.

15. From the functionalist perspective, the family evolves in both form and function in response to changes in the
    A. nature of interaction among members.
    B. conflicts between the workers and owners in society.
    C. larger social environment.
    D. values of the younger generation.

16. The overall poverty rate of black families is
    A. about the same as that of white families.
    B. three times higher than that of white families.
    C. five times higher than that of white families.
    D. less than that of white families only because the white population is larger.

17. Which of the following is NOT a function of family?
    A. Allows humans to reproduce with full social approval
    B. Helps provide physical survival
    C. Prepares people to work in an industrialized economy
    D. Introduces children into their social placement in society

18. Which of the families shown in the video program on family expressed the importance of emotional support?
    A. Elaine and David Starr and their children
    B. Carla Cargile, the single parent, and her children
    C. Louise Young and Vivian Armstrong, the gay couple
    D. All the families

19. Which of the following is NOT a way that families can become more functional?
    A. Acknowledging that each member is an important part of the family
    B. Parents being strong and not expressing their emotions openly
    C. Distinguishing clearly between who are the parents and who are the kids
    D. Modeling humanness, parents not always doing things perfectly

20. Which of the following is a way that families have changed historically?
    A. Modern families are the unit of production.
    B. Nineteenth-century families were the primary unit of consumption.
    C. Modern families are more and more the primary source of socialization.
    D. Step-wise leaving of the family occurs with first the man leaving to go to work, then the woman.

ANSWER KEY

The following provides the answers and references for the practice test questions. Objectives are referenced using the following abbreviations: T = text, R = Telecourse Guide Reading, and V = Video.

Answers	Lesson Goals	Objectives	References
1. B	1	T1	Kornblum, p. 501, Fig. 16.1
2. B	1	T2	Kornblum, p. 502
3. B	1	T3	Kornblum, p. 503
4. B	1	T4	Kornblum, p. 504
5. D	1	T5	Kornblum, p. 506
6. C	1	T6	Kornblum, p. 505
7. B	2	T7	Kornblum, p. 518
8. D	2	T8	Kornblum, p. 518
9. B	2	T9	Kornblum, pp. 519-520
10. B	2	T10	Kornblum, p. 520
11. C	2	T11	Kornblum, p. 521, Box 16.2
12. C	2	T12	Kornblum, pp. 522-523
13. B	2	T13	Kornblum, p. 523
14. D	3	T14	Kornblum, p. 509
15. C	3	T15	Kornblum, p. 510
16. B	3	T16	Kornblum, p. 510, Fig. 16.7
17. C	4	V1	Video
18. D	4	V2	Video
19. B	4	V3	Video
20. D	4	V4	Video

Notes and Assignments:

Lesson 20

Economic Systems

LESSON ASSIGNMENT

Review the following assignment in order to schedule your time appropriately. Pay careful attention; the titles and numbers of the textbook chapter, the telecourse guide, and the video program may be different from one another.

Text:

Kornblum, *Sociology in a Changing World*,
Chapter 19, "Economic Institutions," pp. 606-639.

Reading:

There is no Reading for this lesson.

Video:

"Economic Systems,"
from the series, *The Sociological Imagination*.

OVERVIEW

The call to assist Daren came to Michael from a neighbor. She begged Michael to talk to his old friend, whom she believed was on the verge of an emotional collapse. When Michael arrived, Daren was sitting on his bed, insisting that he was about to cram his tightened fists through the wall.

Daren had lost his job with the local supermarket when it closed four months ago. In almost a whimper, Daren said that he'd worked for twenty years, since he was fifteen, and he knew he'd feel all right about himself if he could just find a job.

Michael assured Daren that he should not blame himself for his situation, that Daren had been a valued and loved member of the staff at the store. As Michael said, the business had closed because of impersonal financial and market forces beyond any individual's control.

Michael also reminded Daren that Daren had been the one who had warned everyone else that the people in the holding company, which was acquiring the supermarket chain through a leveraged buy-out, were ignorant of the grocery business. Daren had been the one complaining about the new owner's lack of attention to equipment maintenance and to proper personnel support in the evenings when stocking and marketing touch-ups had to be done.

Unfortunately, Michael's lesson on the sociology of economics comforted his friend very little. Daren could not see the connection between his personal and family troubles and the macro-economic forces at work in his society, in his community, and in his life. Therefore, he blamed himself for his difficulties.

The video program for this lesson explores the relationship between individuals, families, and economic institutions. You see a man who lost his job and fortune in the oil bust of the late 1980s—something no one person could control.

The lesson also examines how families influence the economy—both as consumers of good and services and as socializing agencies teaching children the values and norms they need for themselves and for the healthy functioning of the economy.

Our goal in this lesson is to develop your sociological imagination about economic systems and institutions, about economic markets, and about the interaction of political systems with those economies. What is the relationship between workers, managers, and corporations? How does the individual get lost in the corporation? What is "alienation" as it applies to the workplace? How is the sociologist's perspective on work and economy different from the economist's?

You will discover that sociologists have some interesting and important things to say about the subject of economy, which frequently is esoteric and incomprehensible to most people. The insights you gain in this lesson will help you to understand one of the most important forces in your life.

Michael tried to help Daren avoid needlessly blaming himself for things that were not private troubles but public and social issues. This new level of awareness should help you to avoid the pitfalls Daren experienced. It also will put you in a better position to make enlightened and positive choices for yourself.

LESSON GOALS

Upon completing this lesson, you should be able to:

1. Analyze the sociological approach to economic institutions, and analyze the relationships among these realities: markets, marketplaces, technology, and multinational corporations.

2. Analyze different economic systems and evaluate the concept of postindustrial society and the changing social contract.

3. Analyze the relationship between workers, managers, and corporations.

4. Analyze the relationship between families and different types of economic systems -including capitalist, socialist, and social democratic. Examine the positive and negative effects of economic change on people.

TEXTBOOK OBJECTIVES

The following textbook objectives are designed to help you get the most from the text. Review them, then read the assignment. You may want to write notes to reinforce what you have learned.

Text: Kornblum, *Sociology in a Changing World*, Chapter 19.

1. Describe the sociological approaches and the major subjects of sociological research on economic institutions, including the ideas of Gary S. Becker and Michael Hechter.

2. Differentiate among these concepts: markets, marketplaces, and market economies, and describe the meaning of contracts.

3. Explain and exemplify the meaning and importance of technology, especially as it relates to changes in economic systems.

4. Define "multinationals."

---

5. Differentiate between mercantilism and laissez-faire capitalism. Explain the importance of private property for capitalism.

6. Define socialism and explain what motivated those who wished to replace capitalism with socialism. Identify the weaknesses of command economies.

7. Describe democratic socialism, where it is practiced, and what its different forms are. Identify nations in which democratic socialism is practiced.

8. Describe welfare capitalism and the changes that have taken place in public attitudes about it.

9. Identify the changes and problems that signal America's becoming a postindustrial society, including changes in the distribution of jobs and the social contract.

10. Identify the major trends in the economics of industrialized nations that are causing workplace conflict.

11. Define the various meanings of the term "alienation" and explain the conclusions of Elton Mayo's research.

12. Describe both the increasing power of the corporation relative to the individual and the implications and consequences of the dominance of corporate actors over individuals.

13. Describe the conflicts that occur at work and explain, exemplify, and evaluate Marxian theory about work conflict.

VIDEO OBJECTIVES

The following video objectives are designed to help you get the most from the video segment of this lesson. Review them, then watch the video. You may want to write notes to reinforce what you have learned.

Video: "Economic Systems"

1. Contrast the economic production of families before and after industrialization began. Describe how family roles changed during the industrial revolution.

2. Explain how economic decisions are made in capitalist, socialist, and social democratic systems.

3. Describe the ways that teaching cultural values within the family affects the economy. Be able to recognize examples.

4. Identify how social changes have affected the economy. Recognize examples in the video program.

RELATED ACTIVITIES

These activities may be used by your instructor as written assignments or as discussion topics. They may also be included as essay questions on your tests.

1. Identify a multinational corporation, and briefly describe the nature of its business. In which countries does this corporation employ personnel and own capital equipment or buildings? In which countries does it sell its product or service?

2. After reviewing the meanings of the term "alienation," explain the definitions of the term to someone you know and ask that person to describe any situations or periods of time when he or she felt a feeling of alienation. Describe and summarize the response. (If you prefer, you can respond to this exercise yourself.)

3. Describe conflicts at your place of work, or at the workplace of a friend or classmate, that are based on class and status differences. Give some specific instances of these conflicts.

4. How old was each member of two or three generations of your family when each began working for a salary? Did the fact that you or others earned your/their own income affect the person's role or relationships in the family? How?

PRACTICE TEST

The following items will help you evaluate your understanding of this lesson. Use the answer key at the end of the lesson to check your answers or to locate material related to each question.

Multiple-Choice

Select the one choice that best answers the question.

1. Michael Hechter concluded that the larger the number of individuals in an organization, the more the organization will be required to
   A. invest its resources in short-term economic gain rather than long-term research and development.
   B. spread and equalize its authority to various levels of organization.
   C. use changing conditions as the basis for improvising means of production and distribution of resources within the organization.
   D. create positions that monitor its members propensity to "do their own thing."

2. The sociological approach to economic institutions is
   A. identical with economists' approach to them.
   B. focused primarily on the medium of exchange used in the distribution of goods and services.
   C. concerned with showing how the norms of different cultures affect economic choices.
   D. based on the assumption that amounts of money or talent have no relationship with choice of values or activities.

3. An agreement in which a seller agrees to supply a particular item and a buyer agrees to pay for it is a
   A. business.
   B. concordat.
   C. concurrence.
   D. contract.

4. Financial and accounting systems that led to the creation of new economic institutions, such as banks, are an example of the influence of
   A. material aspects of culture on the market.
   B. technologies on the new world system.
   C. political power on the development of economic systems.
   D. value of individualism on the market.

5. Economic enterprises that have headquarters in one country and conduct business activities in one or more other countries are called
   A. monopolies.
   B. oligopolies.
   C. multinationals.
   D. corporations.

6. The economic philosophy that held that a nation's wealth could be measured by the amount of gold or other precious metals held by the royal court is known as
   A. laissez-faire capitalism.
   B. mercantilism.
   C. gold-based economy.
   D. metallurgicalism.

7. Socialism was motivated by
   A. a belief that individuals cannot make decisions for themselves.
   B. the horror at the atrocious living conditions caused by industrialism.
   C. a hatred of religion.
   D. large monopolies that wanted even more power.

8. In Yugoslav factories and firms, democratically elected workers' councils make basic managerial decisions, including those that determine who should be hired to manage the firm and how its profits should be invested.

   The Yugoslav example describes a variation of which form of economic philosophy?
   A. Democratic socialism
   B. Welfare capitalism
   C. Socialism
   D. Laissez-faire capitalism

9. Which of the following is NOT a tenet of welfare capitalism?
   A. Markets should have a role in what goods and services will be produced and how.
   B. Government should regulate economic competition by attempting to prevent control of markets by one or a few firms.
   C. Government should allow the market to operate unencumbered by regulation or other interference.
   D. The state should invest in the society's human resources, such as education and health care.

10. Which of the following is a sign of the arrival of postindustrial society?
    A. Centrality of manufacturing is emphasized as the driving force of the economy.
    B. “Intellectual technology” based on information arises alongside machine technology.
    C. Importance of services, especially those employing professional and technical workers, is decreased.
    D. Agricultural jobs increase.

11. Which of the following is NOT a major trend in the economics of industrialized nations that are causing workplace conflict and employee alienation?
    A. Increases in the demand for part-time workers
    B. The trend toward contracting for services by freelance workers
    C. Decreases in the willingness of employers to pay health care benefits
    D. Efforts to encourage employees to join labor unions

12. Karl Marx used which of the following terms to identify the gap between workers and managers?
    A. Capitalist Tension Phenomenon
    B. Hawthorne Effect
    C. Alienation
    D. Postindustrial Split

13. James S. Coleman concluded that one of the effects of the dominance of corporate actors is that
    A. corporations are more responsive to workers and their concerns.
    B. corporations have become vastly more efficient due to the ability to control workers’ lives.
    C. individuals have access to much more information than do the corporate actors, but the latter have more power.
    D. individuals are at a disadvantage in dealings with such actors.

14. Which of the following best describes a generalization of industrial sociologists who take a conflict perspective?
    A. Workers who are permitted to give orders to higher-status employees experience less tension and stress.
    B. Workers limit their output by "making out."
    C. Workers do better when they are given more attention.
    D. Workers will directly challenge the capitalist system itself, thus acquiring desired working conditions and salaries.

15. Before the industrial revolution, families involved in the economic activity of production
    A. were very small and centered around religion.
    B. had more freedom in the choice of roles they would play.
    C. experienced internal conflict due to economic competition.
    D. worked together as a cooperative unit.

16. In which economic system did people observe inequality in the marketplace and take measures to ensure that everyone benefited from the goods and services of the society?
    A. Capitalist
    B. Market
    C. Socialist
    D. Free enterprise

17. Which of the following best illustrates the impact of family and culture on the economic system?
    A. The Randles have seen a television commercial and are discussing whether to buy a new car.
    B. Tasha has just lost her job and is wondering how she is going to break the news to her family.
    C. Phillip's boss has just told him that he has not received the promotion for which he had applied.
    D. Fantasia and William are teaching their children the importance of working hard and saving money.

18. Which of the following best illustrates a way that social changes have affected the economy?
    A. Aging population creating demands for new kinds of housing
    B. Inflation decreasing discretionary income for families
    C. Corporations increasing contributions to legislators
    D. Bankruptcies causing people to lose their jobs

ANSWER KEY

The following provides the answers and references for the practice test questions. Objectives are referenced using the following abbreviations: T = text, R = Telecourse Guide Reading, and V = Video.

Answers	Lesson Goals	Objectives	References
1. D	1	T1	Kornblum, p. 609
2. C	1	T1	Kornblum, p. 608
3. D	1	T2	Kornblum, p. 610
4. B	1	T3	Kornblum, p. 612
5. C	1	T4	Kornblum, p. 613
6. B	2	T5	Kornblum, p. 615
7. B	2	T6	Kornblum, p. 616
8. A	2	T7	Kornblum, pp. 617-618
9. C	2	T8	Kornblum, p. 618
10. B	2	T9	Kornblum, p. 619
11. D	3	T10	Kornblum, pp. 622-623
12. C	3	T11	Kornblum, p. 623
13. D	3	T12	Kornblum, p. 623
14. B	3	T13	Kornblum, p. 626
15. D	4	V1	Video
16. C	4	V2	Video
17. D	4	V3	Video
18. A	4	V4	Video

Notes and Assignments:

Lesson 21

Religion

LESSON ASSIGNMENT

Review the following assignment in order to schedule your time appropriately. Pay careful attention; the titles and numbers of the textbook chapter, the telecourse guide, and the video program may be different from one another.

Text:

Kornblum, *Sociology in a Changing World*,
Chapter 17, "Religion," pp. 532-567.

Reading:

There is no Reading for this lesson.

Video:

"Religion,"
from the series, *The Sociological Imagination.*

OVERVIEW

Dominic could hardly believe his senses. Magnificent gold vessels reflected the light through a haze of wonderfully aromatic incense. The altar held banks of tall candles, glimmering and twinkling like a cluster of stars. Brightly colored vestments glittered with silver lacework as they flowed around their wearers. The great organ music, the bells, the chants, the common prayers delighted his ears. A wondrous spectacle it was indeed.

"It's like a laser show at the rock concert!" Dominic cried out to his mom, kneeling next to him.

"Shh, Dom, don't be disrespectful," she admonished him. "This is a sacred moment."

"What's sacred mean, Mom?"

"Holy. You know, like God. Like when we pray at bedtime—serene and, well, kind of mystical," she whispered.

After the ceremony, standing on the church steps, Dominic enjoyed the hugs he got and the nice people who seemed to enjoy each other so much. He heard their glowing comments about the missionary's inspirational sermon. He hadn't understood most of it, but he was sure it too was really sacred and holy.

Religion was starting to enchant this boy. Even though he was just a child, he knew he was part of something beyond himself, larger even than his family. As he grew older and participated in the church choir and youth groups, he developed an even stronger attachment, a sense of belonging to his church. His belief grew stronger, too. And the everyday routines of his life held more meaning, for they were touched by his faith that God was with him everywhere.

This lesson introduces you to religion as a social institution, which is a different concept from what you know about specific religious institutions. So our first question is: Does this mother's understanding of the sacred match the sociological explanation?

The ceremony that so captivated Dominic was a religious practice we call a ritual. That ritual did not just appear out of thin air. It has a history, as does the religion in which it occurs. How and why have religions formed, from the sociological and historical perspectives?

Dominic's sensory perceptions, even his uncertain awareness of the meaning and value of the missionary's sermon, sparked his first interest in religious belief. What other varieties of religious belief can we categorize?

Little Dominic's excitement and awe were the initial stirrings of what sociologists refer to as religiosity. His religiosity would both proceed from and affect his attachment to his religious community and that particular religious organization. Is organization necessary for religion? How does it develop, and how is it related to the society in which it exists?

The video program examines the functions of religion, both for society and its groups and for individuals. You see and hear practitioners of some of the major religions share their ideas about these functions, based on their own experiences.

Have the influence and growth of religion been constants in human social history? If not, what social and historical events have affected its development? All of these questions are addressed in this lesson, which should give you some new and different insights into religion. Many students of sociology have *experiential* knowledge of religion, sometimes intensive experience. This lesson helps you to develop your *sociological* imagination on the subject.

LESSON GOALS

Upon completing this lesson, you should be able to:

1. Analyze religion, different forms of religious beliefs, and the place of religion in society.

2. Analyze religions as social structures, especially how they change and relate to each other.

3. Analyze religiosity and contemporary trends in religion in the United States.

4. Analyze the functions of religion, using several major religions as examples. Analyze how religion reflects and influences the society and culture in which it exists.

TEXTBOOK OBJECTIVES

The following textbook objectives are designed to help you get the most from the text. Review them, then read the assignment. You may want to write notes to reinforce what you have learned.

Text: Kornblum, *Sociology in a Changing World*, Chapter 17.

1. Explain the meaning of religion and rituals, and differentiate between the sacred and the profane. Identify the type of society in which religion developed as a fully differentiated institution.

2. Define secularization and explain what brought it about. Explain whether or not the importance of religion in public life and, generally, in individuals' lives have diminished.

3. Differentiate between simple supernaturalism and animism, and be able to recognize examples of each.

4. Define theism and the different forms it takes. Describe how Judaism, Christianity, Islam, and Hinduism are theistic religions.

5. Describe the role of religion in bringing about or thwarting social change. Differentiate between Karl Marx and Max Weber's views of the relationship of religion to society and culture.

6. Differentiate among church, sect, and denomination as forms of religious organization.

7. Describe cults and how they are related to the major religions and sects.

8. Explain the views of Max Weber and H. Richard Niebuhr about the relationship of churches and sects to social class. Describe what sects and cults do for individuals, and how they are related to the more established religious organizations.

9. Explain what attracts people to cults. Describe the conflict between individual rights and some cults, and give an example.

10. Describe the role of the black church in America.

11. Define religiosity. Identify the measures of religiosity, and what they tell us about the secularization hypothesis.

12. Define fundamentalism.

VIDEO OBJECTIVES

The following video objectives are designed to help you get the most from the video segment of this lesson. Review them, then watch the video. You may want to write notes to reinforce what you have learned.

Video: "Religion"

1. Define religion and rituals. Describe the importance of rituals and the significance of "the sacred" in religions.

2. Explain the functions that religion fulfills for the personality, the individual. Explain each of the social functions of religion.

3. Explain how religion helps to maintain the status quo in a society. Describe how religion provides social control.

4. The comments of the rabbi and the Muslim leader illustrate the social cohesion function of religion. Explain how their religions provide social cohesion. According to the Muslim leader, explain how his religion contributes to social control. Be able to recognize examples.

5. Explain the legitimizing function of religion. First, show how religion can be a repressive force, perpetuating inequality and racism. Then show how religion provides a source of conflict within society. Finally, show how religion can bring about needed reforms. Be able to recognize examples.

RELATED ACTIVITIES

These activities may be used by your instructor as written assignments or as discussion topics. They may also be included as essay questions on your tests.

1. Many religions have a written record of their "set of coherent answers." Pick one of those religions and describe where these answers are found. Explain how they solve some of the "dilemmas of human existence" for their believers. (You need not pick a religion with whose answers you happen to agree.)

2. Describe at least two rituals of a particular religion and the shared meanings connected with those rituals.

3. Do you agree with Karl Marx's or Max Weber's view of the relationship of religion to society and culture? What evidence do you have to support your opinion? Cite at least three examples.

4. Identify an actual church, a sect, or a denomination. Explain how it fits its definition, illustrating each part of the definition with some aspect of the religious organization.

5. What are some norms or sanctions of a specific religion not mentioned in the video program that help to maintain social control within society? How do these encourage members and others in the community to obey the laws or other norms of the larger community? In a short essay, identify the religion and at least three applicable norms or sanctions.

6. Describe at least two instances in which religion has either generated or thwarted social change. Explain what the change is or was and how the religion affected it.

PRACTICE TEST

The following items will help you evaluate your understanding of this lesson. Use the answer key at the end of the lesson to check your answers or to locate material related to each question.

Multiple-Choice

Select the one choice that best answers the question.

1. Formal patterns of activity that symbolically express a set of shared meanings that may be sacred refers to a definition of
   A. religion.
   B. symbols.
   C. customs.
   D. rituals.

2. Which of the following results in a respect for values of utility and concern for our present welfare on earth rather than of sacredness alone?
   A. Religiosity
   B. Cult awareness
   C. Secularization
   D. Spiritual awakening

3. Individuals who believe that the waters and mountains, the plants and plains are inhabited by the gods have a belief known as
   A. theism.
   B. abstract idealism.
   C. monotheism.
   D. animism.

4. Among the first people to evolve a monotheistic religion were the ancient
   A. Greeks.
   B. Egyptians.
   C. Romans.
   D. Hebrews.

5. In Brazil, the Catholic church's support for the vast numbers of urban and rural poor who seek social justice and equitable economic development is an example of
   A. civil religion.
   B. religion bringing about social change.
   C. economic forces shaping religion.
   D. secularization of society.

6. A religious organization that has strong ties to the larger society is known as a
   A. sect.
   B. cult.
   C. denomination.
   D. church.

7. Most major religions began as seemingly insignificant
   A. cults.
   B. churches.
   C. denominations.
   D. civil religions.

8. According to H. Richard Niebuhr, sects, as they achieve some success, become more
   A. likely to mobilize their membership to oppose oppression of the poor.
   B. like churches and begin to justify existing systems of social stratification.
   C. concerned with justifying the existence of the sect and defending against outside interference.
   D. strongly attached to the ideals that originally inspired the sect.

9. The ongoing conflict between the right of individuals to belong to cults and the efforts to protect them from cult leaders who place themselves above morality and the law is represented by
   A. Protestant denominations outside the mainstream.
   B. followers of Heaven's Gate.
   C. Krishna consciousness.
   D. practitioners of Buddhism.

10. Which of the following is NOT true of the role that churches play in the black community?
    A. They hastened the arrival of secularization in American society because of the emphasis on caring for the welfare of individual members.
    B. They increased economic cooperation.
    C. They assisted in the building of educational institutions.
    D. They became a center of political life.

11. Which of the following is NOT a measure of religiosity?
    A. Strictness of morality
    B. Belief in God
    C. Church or temple attendance
    D. Belief in life after death

12. Religious believers who are devoted to strict observance of ritual and doctrine are called
    A. theists.
    B. fundamentalists.
    C. abstract idealists.
    D. empiricists.

13. Actions that people perform, either privately or in communities, that reaffirm commitment to myth, beliefs, and communities are called
    A. supernaturalism.
    B. rituals.
    C. morality.
    D. dogmas.

14. When people meet in their communities, establish social relationships, and give each other mutual aid, which of the following functions of religion are they fulfilling?
    A. Social control
    B. Social reproduction
    C. Social variability
    D. Social cohesion

15. Which of the following is NOT a way that religion provides for social control?
    A. By gathering together like-minded people in a ritual setting to reaffirm their views of the world
    B. By providing guides for behavior that organize the cosmos between the sacred and the profane
    C. By organizing very real sanctions
    D. By reproducing the values and norms of the existing economic system

16. Which of the following best illustrates the social-control function of religion described by the Muslim leader in the video program on religion?
    A. Muslims should strive for individual salvation.
    B. Muslim speakers should become as eloquent as possible in expressing the beliefs of the Koran.
    C. Muslim people should not show emotion in their attempts to carry out the moral teachings of their religion.
    D. Muslim believers should accept that the biggest Jihad is to fight with themselves, to control themselves.

17. An example of the legitimizing function of religions is
    A. providing a world view for those who supported slavery, thus constraining those who wanted to solve that social problem.
    B. funding the civil rights movement.
    C. repressing the women's movement.
    D. creating conflicts essentially designed to maintain the dominance of one nation over another.

ANSWER KEY

The following provides the answers and references for the practice test questions. Objectives are referenced using the following abbreviations: T = text, R = Telecourse Guide Reading, and V = Video.

Answers	Lesson Goals	Objectives	References
1. D	1	T1	Kornblum, p. 534
2. C	1	T2	Kornblum, p. 536
3. D	1	T3	Kornblum, pp. 538, 539
4. D	1	T4	Kornblum, p. 540
5. B	1	T5	Kornblum, p. 542
6. D	2	T6	Kornblum, p. 543
7. A	2	T7	Kornblum, p. 543
8. B	2	T8	Kornblum, p. 544
9. B	2	T9	Kornblum, p. 545
10. A	2	T10	Kornblum, p. 550, Box 17.1
11. A	3	T11	Kornblum, p. 551
12. B	3	T12	Kornblum, p. 555
13. B	4	V1	Video
14. D	4	V2	Video
15. D	4	V3	Video
16. D	4	V4	Video
17. A	4	V5	Video

Notes and Assignments:

Lesson 22

Mass Media

LESSON ASSIGNMENT

Review the following assignment in order to schedule your time appropriately. Pay careful attention; the titles and numbers of the textbook chapter, the telecourse guide, and the video program may be different from one another.

Text:

Kornblum, *Sociology in a Changing World*,
Chapter 18, "Education and Communications Media," pp. 589-605.

Reading:

There is no Reading for this lesson.

Video:

"Mass Media,"
from the series, *The Sociological Imagination.*

OVERVIEW

The rebels approached the television station, their hopes of success running high. They knew if they could capture the station and begin sending sound and pictures indicating the overthrow of the regime, their victory was assured. They could then tell *their* side of the story, instead of the pernicious propaganda broadcast by the present rulers. Although most of these "rebels" were peasants and poor urban dwellers, they were sophisticated enough to understand the power of mass media.

The media have tremendous influence at several different levels. On a cultural level, they shape values and attitudes. They can perpetuate ideologies and can create wishes and desires. The dazzling success of the Polaroid Company was due to the effectiveness of its advertising in creating a "need" for instantly developed photos. On a sociopolitical level, governments can be toppled if they lose control of their media institutions. A striking example is the change in attitude of many staunch Party members toward East German communism, after they learned over television about the abuse perpetrated by their former leaders. When the "old guard" controlled the media, the opulent lifestyles of these leaders went unreported, and therefore unnoticed by the masses who toiled and struggled daily to make ends meet in a failing economy.

At the sociopsychological level, there is considerable controversy about whether television violence has a deleterious effect on the behavior and attitudes of children and teenagers. In this lesson you find out more about the effects of mass media on individuals and on society. Some of your assumptions about this issue may be challenged by the sociological evidence.

The mass media comprise an important social institution in modern societies, one that serves many positive functions. They support economic development; they make people aware of events and realities far beyond the bounds of their neighborhoods and communities; they can be used to further democracy and positive development of governmental and other social institutions.

What is the relationship of the institution of mass media to the economy? Are newspapers, magazines, television, and radio merely puppets of big business and profit machines? And what is the relationship of mass media to political institutions? As alluded to above, mass media can be extremely powerful instruments in the hands of politicians and their media experts; we already see the use of mass media in political campaigns transforming the political landscape.

These issues are important to every member of society. They affect us at all levels of our experience.

LESSON GOALS

Upon completing this lesson, you should be able to:

1. Analyze mass media's importance and influence on individuals and society.

2. Analyze the power and limits of mass media.

3. Analyze the influence of mass media on us as individuals, as well as on our society and culture.

TEXTBOOK OBJECTIVES

The following textbook objectives are designed to help you get the most from the text. Review them, then read the assignment. You may want to write notes to reinforce what you have learned.

Text: Kornblum, *Sociology in a Changing World*, Chapter 18.

1. Explain the meaning and importance of the communications media. Explain and exemplify how norms that establish media freedom can conflict with norms that are designed to control media.

2. Describe the effect mass media have had on American culture and how patterns of media consumption have changed over time.

3. Discuss social-scientific evidence about the effects of television violence on human behavior.

4. Describe the importance of electronic media and the relationship between electronic media, advertisers, and business.

5. Explain how television is related to political institutions in the United States and internationally. Give examples.

6. Explain how media power is limited by both technology and the nature of communication itself.

VIDEO OBJECTIVES

The following video objectives are designed to help you get the most from the video segment of this lesson. Review them, then watch the video. You may want to write notes to reinforce what you have learned.

Video: "Mass Media"

1. Explain why media are important to people in a modern society. Identify the benefits people get from listening to music on the radio. Explain how disk jockeys and radio program mangers influence a culture.

2. Identify the benefits and drawbacks of television, including advertising.

3. Outline the relationship between reporters and governmental sources of information. Describe the power and influence of television on people. Discuss whether or not you believe mass media manage the status quo.

RELATED ACTIVITIES

These activities may be used by your instructor as written assignments or as discussion topics. They may also be included as essay questions on your tests.

1. Survey your friends and relatives, asking them where they get the news that is important to them. Do they rely solely on interpersonal communications, or do they use the newspapers, television, and radio? How frequently do they use these various media during a given week? Report your findings.

2. Think about one television show with violent, aggressive behavior and one without it. Write a brief narrative describing your feelings after each show you watched.

3. Make a list of your family members and closest friends. Based on what you know, next to each name write the kind of music that person likes to listen to. What does this tell you about the influence radio and music have on people's lives?

4. What kind of music do you most enjoy? How does the mass media portray people who listen to this type of music? Do you "fit" the portrayal?

5. Do you agree with Professor Todd Gitlin about television's power and influence on people? Explain your reasons.

PRACTICE TEST

The following items will help you evaluate your understanding of this lesson. Use the answer key at the end of the lesson to check your answers or to locate material related to each question.

Multiple-Choice

Select the one choice that best answers the question.

1. Which of the following is NOT true of communications media?
   A. They use symbols to tell us something about ourselves and our environment.
   B. They are synonymous with formal education.
   C. They include print, movies, radio, and television.
   D. They are institutions that specialize in communicating information, images, and values about ourselves, our communities, and our society.

2. Which of the following best describes the influence of mass media on American culture?
   A. Sociological research data overwhelmingly support the idea of mass culture.
   B. Newspapers remain an important source of news, information, and entertainment for American adults.
   C. Radio does not compete well with television for people's attention.
   D. The average American household has its television turned on for almost 15 hours per week.

3. The effects of television viewing, especially viewing violence,
   A. are designed to affect most viewers in similar ways.
   B. cause most viewers to be tempted to act out that violence.
   C. seem to depend on the viewer's emotional condition.
   D. are not related to aggressive behavior in people.

4. Today the dominant communications institutions in the United States are
   A. newspapers.
   B. magazines.
   C. electronic media.
   D. book publishers.

5. In contemporary societies where totalitarian regimes try to control the minds of citizens,
   A. media take a particularly strong role in providing alternative views of government.
   B. military institutions are separated from the media.
   C. total control over the media is a major goal of the state.
   D. expression of divergent views occurs through the arts.

6. Which of the following does NOT represent the limits on media power?
   A. "Opinion leaders" exert a powerful influence by means of oral, interpersonal communication.
   B. Video cassette equipment makes possible a wide range of choices for the consumer.
   C. Researchers have found a direct link between persuasive messages and actual behavior.
   D. Cable television with its diversity has expanded the choices of media consumers.

7. Mass media are important to people because they
   A. tell people who they are, what they should think, and what values are appropriate.
   B. help people develop critical thinking skills.
   C. reinforce the diversity of American culture and provide a voice for minorities and other oppressed people.
   D. help people question the directions taken by their political system.

8. One of the detriments or negative effects of television is that it
   A. promotes the growth of fringe groups within the political system.
   B. sometimes opposes religious values that are important.
   C. invites people to think of themselves as incomplete and capable of being satisfied by the purchase of goods and services.
   D. depresses people with a barrage of news stories about social problems and other evils in the world.

9. Describe the relationship between reporters and their government sources.
   A. Symbiotic
   B. Antagonistic
   C. Neutral
   D. Competitive

ANSWER KEY

The following provides the answers and references for the practice test questions. Objectives are referenced using the following abbreviations: T = text, R = Telecourse Guide Reading, and V = Video.

Answers	Lesson Goals	Objectives	References
1. B	1	T1	Kornblum, p. 589
2. B	1	T2	Kornblum, p. 591, Table 18.6, p. 593
3. C	1	T3	Kornblum, p. 595
4. C	2	T4	Kornblum, p. 590
5. C	2	T5	Kornblum, p. 595
6. C	2	T6	Kornblum, p. 595
7. A	3	V1	Video
8. C	3	V2	Video
9. A	3	V3	Video

Notes and Assignments:

Lesson 23

Political Systems

LESSON ASSIGNMENT

Review the following assignment in order to schedule your time appropriately. Pay careful attention; the titles and numbers of the textbook chapter, the telecourse guide, and the video program may be different from one another.

Text:

Kornblum, *Sociology in a Changing World*,
Chapter 20, Politics and Political Institutions," pp. 640-669.

Reading:

There is no Reading for this lesson.

Video:

"Political Systems,"
from the series, *The Sociological Imagination*.

OVERVIEW

"Scott, get out of that refrigerator! We'll be eating in just an hour. I don't want you spoiling your appetite."

"Oh, Mom, I'm *real* hungry. I'll still be hungry after I eat an ice cream bar. I've got a B-I-G space, right here." The ten-year-old grinned winningly as he pointed to his lower abdomen.

But when he saw the frown on his mother's face, he reluctantly closed the freezer door and began to pout.

Just then, Ginny, Scott's seventeen-year-old sister, ran into the house, announcing that a television crew had cornered the mayor at the supermarket. They were asking him about the city council's threat to impeach him for fraudulent use of city workers. "Maybe *I'll* run for mayor if they get rid of him," she said.

"I don't think you're old enough yet," said her mother.

"But I've helped the party raise money. And I went house to house passing out flyers for our national candidates last year. Just think of all those envelopes I licked! Besides, this is a democracy, and I'm ready," said Ginny with determination.

This vignette shows politics happening at several levels. Each instance entails the use of power. The interaction between Scott and his mother is a micro-level exercise of power within the family. Does Mom have the authority to order her son the way she did?

If Scott's pouting had induced Mom to relent on the ice cream, would he have successfully exercised power—or influence? Influence is the use of personality to prevail upon others, while power involves some kind of threat, either implied or actual.

Scott tried to influence his mother with his semi-comic argument about the intensity and capacity of his hunger. When that failed, he tried another strategy that sometimes worked: pouting. Scott was bringing to bear on this situation all the personal resources he could think of at the moment to sway, or influence, his mother. But Scott also recognized the implied threat in his mother's frown.

Next, how is it that the duly-elected mayor might be removed from office? Power is not without restraints, and there is only a probability—not a guarantee—that a person will be able to maintain power. At this moment the rightfulness of the mayor's power obviously is in question. What is the difference between power and authority—in this case, the difference between the mayor's power, tenuous at the moment, and his authority, while still an elected official? What is legitimacy?

Ginny is contemplating entering the political arena. She, along with millions of other citizens, has exercised her rights in the past—both by voting and by getting out and actively supporting the party of her choice. Now she is thinking about using the existing political institutions to build a possible future for her and her community. What is a political institution? How do political parties work? In fact, what is a democracy, and how does it differ from other political systems?

This lesson introduces you to these issues. The video program examines various political systems, with a Cuban expatriate talking about autocracy and totalitarianism and a refugee from the former Soviet Union telling his story about political oppression.

Building your awareness of political systems will take you on yet another leg of your journey into your sociological imagination.

LESSON GOALS

Upon completing this lesson, you should be able to:

1. Analyze power and legitimacy, and how they are used by political institutions.

2. Analyze the conflicts that arise over political borders and territories, and analyze how the rights of citizenship vary over time and place.

3. Analyze how political leaders deal with conflicts.

4. Analyze the major perspectives about the way problems of inequality and injustice are addressed in the United States.

5. Analyze the power and functions of the military, and understand the process of military socialization.

6. Analyze the uses of power at various levels, especially at the level of government, and analyze different kinds of government, including autocracy, totalitarianism, and democracy.

TEXTBOOK OBJECTIVES

The following textbook objectives are designed to help you get the most from the text. Review them, then read the assignment. You may want to write notes to reinforce what you have learned.

Text: Kornblum, *Sociology in a Changing World*, Chapter 20.

1. Define power and authority. Differentiate between examples of each.

2. Define political institutions. Cite examples. Differentiate between the views of politics of Karl Marx and Max Weber.

3. Define legitimacy and explain why it is necessary.

4. Differentiate among traditional, charismatic, and legal authority. Be able to recognize examples of each. Identify which type of authority is most adaptable to social change.

5. Describe the nation-state's authority within its borders.

6. Explain the meaning of citizenship, and describe the rights it includes. Differentiate between citizenship in feudal and in modern societies.

7. List and describe Denis and Derbyshire's classifications of political regimes.

8. Describe the view of Seymour Martin Lipset about the central problem of modern politics.

9. Identify how modern societies expect political leaders to use their authority. Explain how the power of political leaders is controlled in modern nations. Describe what a demagogue is. Cite examples of the latter.

10. Describe the power elite model. Identify one criticism of this model.

11. Explain the pluralist model. Describe interest groups and lobbying.

12. Describe the economic power and functions of the military.

13. Explain the interactionist perspective on the military, and outline how military socialization takes place.

VIDEO OBJECTIVES

The following video objectives are designed to help you get the most from the video segment of this lesson. Review them, then watch the video. You may want to write notes to reinforce what you have learned.

Video: "Political Systems"

1. Describe the use of power in families, both historically and today. Describe who holds the power in schools and churches.

2. Explain what an autocracy is. Cite examples of an autocracy as shown in the video program on political systems.

3. Define totalitarianism. Describe the ways totalitarianism influences the daily lives of the people under its rule, using the man who lived in Russia as an example.

4. Define democracy, and be able to recognize its various forms. Describe when and under what conditions people engage in political action. Exemplify systems that have checks and balances for those selected to rule.

RELATED ACTIVITIES

These activities may be used by your instructor as written assignments or as discussion topics. They may also be included as essay questions on your tests.

1. There is a significant difference between power and influence. You exercise influence when you are able to persuade people to do what you want through use of argument or sheer personality. In that instance, the people you wish to influence make their own decisions whether they will follow your wishes, based on their respect or lack of respect for you as a person. With power, you are able to control others because of some threat. The threat may be implied and somewhat hidden, or it may be explicit and external; all that is necessary is the perception that your power is real.

   Describe an instance in which you were able to successfully exercise power over someone—even against his or her resistance. What was the situation, and who was involved? How did the other person resist, and what was the nature of your threat?

2. Describe two typical situations in your family, or in a close relationship, in which politics operates.

3. In any of your groups, have you ever witnessed a case in which a person in authority had the "rightfulness" or legitimacy of his or her power questioned? Such an instance may have involved a religious or civic leader, a boss, or another official. Describe the situation. Who was the authority involved?

What brought about the challenge to his or her legitimacy? What moral arguments or values were cited during the challenge? Who won—and why?

4. What is your opinion about the effectiveness of the political institutions of your society in dealing with problems of inequality and injustice? Cite at least three pieces of evidence to support your view.

5. Would you describe the power structure in your city or town as being a power elite or a pluralist structure? Give arguments to support your view.

6. List at least three interest groups that are involved in lobbying. Describe the goal or purpose of each group, and tell how many members are in each.

7. Interview someone who has served in the military. Using the material about military socialization as your guide, summarize at least three instances in which that person experienced the socialization process within the military institution. What were the formal and informal aspects of the process?

PRACTICE TEST

The following items will help you evaluate your understanding of this lesson. Use the answer key at the end of the lesson to check your answers or to locate material related to each question.

Multiple-Choice

Select the one choice that best answers the question.

1. The ability to control the behavior of others, even against their will, is
   A. social control.
   B. power.
   C. influence.
   D. leadership.

2. The judiciary is an example of a political
   A. system.
   B. society.
   C. culture.
   D. institution.

3. The capacity of a society to engender and maintain the belief that existing political institutions are the most appropriate for the society is called
   A. power.
   B. influence.
   C. legitimacy.
   D. rational choice.

4. Legal authority is legitimated by people's belief in the supremacy of
   A. God's divine will.
   B. the leader's personal gifts.
   C. the law.
   D. traditions and customs.

5. Nation-states claim a legitimate monopoly over
   A. power within their boundaries.
   B. economic processes within the state.
   C. cultural institutions within their boundaries.
   D. use of force within their borders.

6. Which of the following is NOT listed by T.H. Marshall as a right of citizenship?
   A. Civil rights
   B. Political rights
   C. Social rights
   D. Global rights

7. Which of the following types of national regimes is characterized by extreme nationalism leading to intolerance and exclusion of other races and creeds?
   A. Liberal democracies
   B. Communist regimes
   C. Nationalistic socialist regimes
   D. Authoritarian nationalist regimes

8. According to Seymour Martin Lipset, the central problem of modern politics is how a society can incorporate
   A. bureaucratic forms into the political process.
   B. charismatic leadership into mainstream political structures.
   C. increasingly mobile people from the middle classes into the political institutions.
   D. continuous conflict among its members and groups and still maintain social cohesion and state authority.

9. Which of the following is NOT true of the way modern societies expect political leaders to use their authority?
   A. Leaders, both elected and appointed, govern with authority defined by laws.
   B. Leaders are expected to use authority for the good of all rather than for their own benefit.
   C. Leaders have the authority to represent interests of particular groups even though those interests may contradict the needs of the masses.
   D. Leaders are forbidden to use authority in illegitimate ways.

10. Having status immediately below the power elite are the
    A. professional politicians at the middle levels of power.
    B. military generals and top brass who assist the elite.
    C. mid-managers in corporations who take their bidding from that elite.
    D. information experts who provide raw data necessary to maintain power.

11. Interest groups are
    A. advocates of causes that often develop into mainstream political parties.
    B. charitable organizations that sponsor benefits.
    C. publics with the same opinion about social, economic, cultural, or political issues.
    D. specialized organizations that attempt to influence elected and appointed officials on specific issues.

12. The theory that many decision makers are engaged in a process of coalition building and bargaining is known as the
    A. power-elite model.
    B. pluralist model.
    C. democratic model.
    D. participatory model.

13. Even in the United States, where the threat of military coups is not considered great, the military has so much influence on the economy that
    A. military officers are able to achieve promotions illegally.
    B. social control of the military has become difficult.
    C. social organization within the military has become overly complex and therefore unmanageable.
    D. nationwide recessions and depressions can be caused by minor changes in defense budgets.

14. Which group is most interested both in its own survival and in adhering to group norms?
    A. Fighter jocks
    B. Officers
    C. Draftees
    D. Career military

15. Which of the following best describes who has had power in American families historically?
    A. The father has had all the power.
    B. The young child of modern society has had most of the power.
    C. The matriarchal family has had a dominant place in American society.
    D. The child has had dependency upon, as well as reciprocal power with, parents.

16. In an autocracy, power is held by
    A. one person with sole authority.
    B. three branches of government with equal authority.
    C. one person or group, with support by people in key positions.
    D. ruling elite, with no one person having more power than another.

17. The man interviewed in the video program on political systems who told stories about being spied upon by neighbors and about security police searching houses for bibles and any sign of possible anti-government material lived in what kind of political system?
    A. Totalitarianism
    B. Autocracy
    C. Socialist democracy
    D. Oligarchy

18. A form of government in which the rulers rule with the consent of the governed is
    A. an autocracy.
    B. a totalitarian state.
    C. a democracy.
    D. a communistic state.

ANSWER KEY

The following provides the answers and references for the practice test questions. Objectives are referenced using the following abbreviations: T = text, R = Telecourse Guide Reading, and V = Video.

Answers	Lesson Goals	Objectives	References
1. B	1	T1	Kornblum, p. 642
2. D	1	T2	Kornblum, p. 643
3. C	1	T3	Kornblum, p. 644
4. C	1	T4	Kornblum, p. 644
5. D	2	T5	Kornblum, p. 647
6. D	2	T6	Kornblum, p. 649
7. D	2	T7	Kornblum, p. 651
8. D	3	T8	Kornblum, p. 649
9. C	3	T9	Kornblum, pp. 649-650
10. A	4	T10	Kornblum, p. 656
11. D	4	T11	Kornblum, p. 656
12. B	4	T11	Kornblum, p. 656
13. B	5	T12	Kornblum, p. 660
14. C	5	T13	Kornblum, p. 660
15. D	6	V1	Video
16. C	6	V2	Video
17. A	6	V3	Video
18. C	6	V4	Video

Notes and Assignments:

Lesson 24

Science and Technology

LESSON ASSIGNMENT

Review the following assignment in order to schedule your time appropriately. Pay careful attention; the titles and numbers of the textbook chapter, the telecourse guide, and the video program may be different from one another.

Text:

Kornblum, *Sociology in a Changing World*,
Chapter 21, "Science, Technology, and the Environment," pp. 670-696.

Reading:

There is no Reading for this lesson.

Video:

"Science and Technology,"
from the series, *The Sociological Imagination*.

OVERVIEW

Late in the night the huge supertanker lumbered through the choppy, frigid waters off the coast of Prince William Sound. Without warning, it came to a full stop. This was no ordinary pause in the journey of the tanker, fully loaded with Alaskan crude oil. Something dreadful was wrong. A call went out to the Coast Guard for help.

Eventually, much more help than a few night-shift Coast Guard operators would be needed. Hundreds of volunteers and professionals would work for months to clean up one of the largest and costliest oil-spill disasters in history. The costs would not be only in dollars, for thousands of wildlife would be killed and a once-pristine coastline, sullied and contaminated.

If this incident was only a once-in-a-lifetime occurrence, it would still be a tragedy. But it was no such thing. In the penultimate decade of the twentieth century, Planet Earth would suffer a recurring pattern of similar accidents involving spills of oil of varying grades and levels of refinement.

Many citizens are frustrated with the seeming inability or unwillingness of industry officials to stop this fouling of the environment. Yet we are uncertain about oil supplies from international sources. Americans, once convinced of the ability of technology to solve almost any problem—and certainly its own problems—are beginning to reappraise that total faith in science and technology.

Scientists, secure in their obligation to bring complete objectivity and the other rigors of the scientific method to their work, tout the reliability and accuracy of their findings. They stand in justifiable pride before the technological innovations that have been the fruits of their labor: everything from the light that illuminates the words of the page you are reading, to the integrated circuits and microprocessors that make up the computer on which this author is writing these words. And yes, even the refining capacity of the petrochemical industry itself is based on that particular kind of knowledge we call science.

Sociologists, too have always understood their field as a science. So they have been careful to apply the rules and methods of science to their investigations and procedures. But sociologists also have occupied themselves with analyzing science itself, because they understand that any human endeavor takes place in a social context.

Scientists of all kinds do what they do within the concentric circles of groups, formal organizations, communities, and societies. Their work is both enabled and constrained by social institutions: the economy, government, education, religion, and even family.

This lesson is about that social context in which science takes place. Through it, you also gain insight into technology, specifically, two related spheres: the environment and medicine.

LESSON GOALS

Upon completing this lesson, you should be able to:

1. Analyze the relationship between science and technology, and analyze the norms and functions of science.

2. Analyze the dimensions of technology, its good and bad effects, and problems in its development.

3. Analyze problems that arise from technological innovation, especially as they pertain to health care and the environment.

4. Analyze the social context in which the social institution of science and technology occurs, and analyze how AIDS research illustrates this social context.

TEXTBOOK OBJECTIVES

The following textbook objectives are designed to help you get the most from the text. Review them, then read the assignment. You may want to write notes to reinforce what you have learned.

Text: Kornblum, *Sociology in a Changing World*, Chapter 21.

1. Define science, and differentiate between pure and applied science.

2. Explain the meaning of technology. Describe the history of the relationship between science and technology.

3. Describe the function of institutions of science. Describe and give examples of universalism and common ownership as norms of science.

4. Explain the meaning of disinterestedness as a norm of science. Give examples.

5. Identify the dimensions of technology.

6. Define and give example of technological dualism.

7. Explain the meaning of and exemplify cultural lag as it applies to our society.

8. Identify the major difference between the two "energy paths," according to Amory Lovins. Be able to recognize examples of both.

9. Define hypertrophy. Explain how it is illustrated in the case of health care. Describe Paul Starr's analysis of the problem with the American health-care system.

10. Identify what medical sociologists study. Describe the critical health-care issues that have arisen as a result of advancing technology. Assess how well equipped health-care institutions are to deal with these issues.

11. Describe and exemplify the "new kind of technological catastrophes."

12. Differentiate between pollution and environmental stress by showing how each can affect our lives. Be able to recognize examples of both.

VIDEO OBJECTIVES

The following video objectives are designed to help you get the most from the video segment of this lesson. Review them, then watch the video. You may want to write notes to reinforce what you have learned.

Video: "Science and Technology"

1. Using the example of research on the AIDS virus, explain how science occurs in a social context. Describe what kind of competition is involved in the research process. Describe the impact of politics and the economy on scientific competition.

2. Explain how religion has affected science. Cite historical examples, including AIDS research and treatment.

3. Explain how the community—for example, pressure groups—has affected science. Be able to recognize examples from AIDS research.

RELATED ACTIVITIES

These activities may be used by your instructor as written assignments or as discussion topics. They may also be included as essay questions on your tests.

1. Write a short essay on the difference between the way science and religion approach problems.
2. Identify and list two examples of each of the dimensions of technology.
3. Identify at least two technological innovations in your home, then describe two positive and two negative effects of each. You may include physical, social, and psychological effects.
4. Describe a modern-day crime or social problem (not mentioned in the textbook) that is a result of cultural lag. Explain how cultural lag is involved.
5. What are the arguments for and against the right to die? With which of these views do you agree and why? Write a brief essay presenting both views about this critical health care issue.

PRACTICE TEST

The following items will help you evaluate your understanding of this lesson. Use the answer key at the end of the lesson to check your answers or to locate material related to each question.

Multiple-Choice

Select the one choice that best answers the question.

1. Science is knowledge that has been
   A. passed down by tradition.
   B. obtained from developing and testing of hypotheses.
   C. used by inventors of technology because of the influence of respected authority.
   D. accepted by the general population as true.

2. The use of tools and knowledge to manipulate the physical environment in order to achieve desired practical goals is referred to as
   A. pure science.
   B. technology.
   C. institutionalized science.
   D. environmental application.

3. One of the consequences of the norm of common ownership is the norm of
   A. universalism.
   B. publication.
   C. disinterestedness.
   D. altruism.

4. That a scientist does not allow the desire for personal gain to influence the reporting and evaluation of results is known as
   A. disinterestedness.
   B. objectivity.
   C. subjectivity.
   D. universalism.

5. Technological tools, instruments, machines, and gadgets that are used in accomplishing a variety of tasks are known as
   A. instrumentalities.
   B. mechanisms of science.
   C. appliances.
   D. apparatus.

6. The safety hazards associated with the use of power tools are an example of
   A. scientific application of technology.
   B. poor application of scientific knowledge.
   C. technological dualism.
   D. cultural particularism.

7. Rapid changes in office technology have meant that many schools are unable to maintain current equipment and up-to-date training procedures.

   This is an example of
   A. the Gaia hypothesis.
   B. cultural lag.
   C. technological assessment.
   D. technological pollution.

8. Rapid expansion of centralized high technologies to increase supplies of energy is referred to as
   A. hard energy paths.
   B. moderate energy paths.
   C. soft energy paths.
   D. direct energy paths.

9. Critics of modern medicine who argue that the American health system is suffering from "hypertrophy" mean that
   A. health care has become too expensive for the average person.
   B. needs of patients are subordinated to needs of providers.
   C. numbers of doctors in the United States are insufficient.
   D. medical advances have not been effective in extending life expectancy.

10. Which of the following is NOT a concern of medical sociologists?
    A. Size of the medical profession
    B. Public policy regarding access to medical care
    C. Ethical issues in the use of life-prolonging technologies
    D. Criteria for deciding what constitutes health

11. According to Charles Perrow, the major cause of technological catastrophes in the modern world is
    A. malfunctioning apparatus.
    B. ineffective safety precautions.
    C. failures of whole systems.
    D. overworked maintenance personnel.

12. Which of the following refers to the effects of society on the natural environment?
    A. Environmental stress
    B. Pollution
    C. Structural disequilibrium
    D. Technological displacement

13. Which of the following is true of the effects of competition on science, especially in the case of AIDS research?
    A. Competition was non-existent among scientists because of the objectivity and rigor in the scientific method.
    B. Competition became very intense.
    C. Competition was hastened because of discovery of the electron microscope.
    D. Competition was lessened by the fact that French scientists could speak English with their American colleagues.

14. Which of the following is NOT an example of scientific research that religion tried to suppress?
    A. The research of Dr. Jonas Salk, who developed the polio vaccine
    B. The work of Copernicus, who explained that the earth rotates on its axis and the planets revolve in orbits
    C. The conclusions of Galileo, who discovered that the earth revolves around the sun
    D. The ideas of Darwin, who formulated the theory of evolution

15. Which of the following best illustrates the effect of the community on science?
    A. Purchase of large quantities of scientific equipment by companies doing research on the AIDS virus
    B. Acknowledgment by universities of scientists who make significant discoveries in the area of AIDS
    C. Changes in the nature of drug testing protocols due to pressure by gay rights groups
    D. Alteration of organizational structures by faculty who believe that science must be maximally efficient

ANSWER KEY

The following provides the answers and references for the practice test questions. Objectives are referenced using the following abbreviations: T = text, R = Telecourse Guide Reading, and V = Video.

Answers	Lesson Goals	Objectives	References
1. B	1	T1	Kornblum, p. 672
2. B	1	T2	Kornblum, p. 672
3. B	1	T3	Kornblum, p. 676
4. A	1	T4	Kornblum, p. 678
5. D	2	T5	Kornblum, p. 679
6. C	2	T6	Kornblum, p. 679
7. B	2	T7	Kornblum, pp. 681-682
8. A	2	T8	Kornblum, p. 683
9. B	3	T9	Kornblum, p. 689
10. A	3	T10	Kornblum, pp. 689-690
11. C	3	T11	Kornblum, pp. 689-690
12. A	3	T12	Kornblum, p. 684
13. B	4	V1	Video
14. A	4	V2	Video
15. C	4	V3	Video

Notes and Assignments:

Lesson 25

Collective Behavior and Social Movements

LESSON ASSIGNMENT

Review the following assignment in order to schedule your time appropriately. Pay careful attention; the titles and numbers of the textbook chapter, the telecourse guide, and the video program may be different from one another.

Text:

Kornblum, *Sociology in a Changing World*,
Chapter 8, "Collective Behavior and Mass Publics," pp. 226-255.

Reading:

There is no Reading for this lesson.

Video:

"Collective Behavior,"
from the series, *The Sociological Imagination*.

OVERVIEW

Tiananmen Square, spring 1989. The trenchant picture of the lone Chinese student intimidating and stopping the tank in its tracks will endure in the minds of historians, sociologists, and all lovers of democracy for years to come. Some believe Tiananmen Square was an incident of collective behavior that was the precursor to the social movements that rocked the communist world in 1989 and the early 1990s.

Collective behavior is a subject of great interest to sociologists. You are now aware of how people act in more structured social entities, such as groups, complex organizations, communities, and societies. But collective behavior refers to less structured behavior and therefore includes an even wider range of actions. Some of those actions are spontaneous; some are motivated by strong emotions. Some are not very well organized, while others are highly organized.

Mass media have helped to make us aware of the importance of collective behavior, especially when that behavior reflects or produces social movements. We have seen and heard the demonstrations mounted by reform movements in Eastern Europe, and we have seen and read about the massive social changes occurring as a result of those movements.

When you write your Social Security number on an application, you probably take for granted that massive bureaucracy whose purpose is to provide some economic benefit to you in your retirement. Yet that institution resulted from a social movement several decades ago—and the benefits that accrue to people who receive Social Security are kept at a relatively good level as a result of another, more recent social movement.

If you have ever been to a festival and become excited by the sights, sounds, and smells around you, you no doubt acted differently than you do at home reading a book. You and the others there were experiencing collective behavior.

Have you ever purchased a "fad" item, such as a piece of clothing, a toy or a game, or gone along with some new trend? If so, your fascination, and how you acted on it, represents yet another form of collective behavior.

In this lesson and in the video program, you see several examples of collective behavior, ranging from simple actions to a social movement. You gain new and deeper insights into collective behavior as a form of social behavior and see how it develops, the many different forms it takes, and how it affects groups, institutions, and whole societies.

LESSON GOALS

Upon completing this lesson, you should be able to:

1. Analyze collective behavior and the diversity of its forms.

2. Analyze social movements, especially how they develop and how they are categorized.

3. Analyze mass publics and public opinion.

4. Analyze the different types of social movements and the methods they use to bring about change.

TEXTBOOK OBJECTIVES

The following textbook objectives are designed to help you get the most from the text. Review them, then read the assignment. You may want to write notes to reinforce what you have learned.

Text: Kornblum, *Sociology in a Changing World*, Chapter 8.

1. Explain the meaning of collective behavior, and describe the range of behaviors it includes.

2. Define a social movement and recognize examples that show the diversity of social movements.

3. Differentiate between a crowd and a mass, and explain the role of emotions in each.

4. Identify the four goal-based categories of social movements, and define an expressive social movement. Recognize an example of each.

5. Differentiate between political and social revolutions, and give an example of each. Explain the "relative deprivation" theory.

6. Identify the stages that many social movements go through. Explain the importance of charisma to social movements, and describe co-optation.

7. Describe and recognize examples of mass publics.

8. Describe public opinion and explain how it is shaped. Recognize examples of behaviors that develop out of public opinion.

VIDEO OBJECTIVES

The following video objectives are designed to help you get the most from the video segment of this lesson. Review them, then watch the video. You may want to write notes to reinforce what you have learned.

Video: “Collective Behavior”

1. Distinguish between collective behavior and social movements. Recognize examples of each. Explain the difference between a radical social movement and a reform movement.

2. Identify the goal of revolutions. Discuss how the American, Bolshevik, French, and Maoist Revolutions were different from each other. Discuss what they all have in common.

3. Identify the conditions that must be present in a society for a reform movement to be effective. Identify four major reform movements in recent American history, and describe the extent to which they reached their goals.

4. Describe how society's awareness of environmental issues has changed over time. Discuss how successful the environmental movement has been. Identify the organizations and activities that have been part of this movement.

RELATED ACTIVITIES

These activities may be used by your instructor as written assignments or as discussion topics. They may also be included as essay questions on your tests.

1. List five specific non-textbook examples of collective behavior, beginning with the least organized and moving to the most organized.

2. Identify a recent or current social movement with a leader who has charisma—as defined by Max Weber. Identify a local or national leader, and describe the special qualities of that leader. Then describe situations in which the leader appears to be inspiring followers. What does the leader say and do? What do the followers do that indicate that they are being motivated?

3. Describe a recent example of mass public behavior. In your description, give some indication of the size of the population involved. What actions, events, and situations surrounded the collective behavior?

4. Identify a current or past radical social movement, and describe the changes in the social system it is trying or tried to bring about.

5. Write an essay depicting and illustrating the ideology of a reform

PRACTICE TEST

The following items will help you evaluate your understanding of this lesson. Use the answer key at the end of the lesson to check your answers or to locate material related to each question.

Multiple-Choice

Select the one choice that best answers the question.

1. Which of the following best describes behavior that is typically referred to as collective behavior?
   A. Regular interactions existing in primary groups
   B. Formalized relationships in bureaucracies and other formal organizations
   C. Dynamic behavior usually seen in collections of odd-numbered groupings in which conflict typically occurs
   D. Spontaneous behavior of people to situations they perceive as uncertain, threatening, or extremely attractive

2. Intentional efforts by groups in a society to create new institutions or reform existing ones are called
   A. social movements.
   B. social organization.
   C. democratic government.
   D. public agendas.

3. A large number of people oriented toward a set of shared symbols or social objects is called a
   A. mass.
   B. crowd.
   C. social group.
   D. society.

4. Which of the following is NOT a category of social movement?
   A. Reformist
   B. Conservative
   C. Reactionary
   D. Formative

5. Transformations in the political structures and leadership of a society that are not accompanied by rearrangement of productive capacities, culture, and stratification systems are known as
   A. political revolutions.
   B. social revolutions.
   C. reformist movements.
   D. reactionary movements.

6. Which of the following is NOT an aspect of the role of charisma in a social movement?
   A. Special qualities of leaders that motivate people to follow
   B. The ability to organize and structure activity logically
   C. Goals and gifts of leaders incorporated into the structure of the movement
   D. Institutions of society taking on goals of movements

7. Which of the following is NOT an example of behavior typical of mass publics?
   A. Panics
   B. Crazes
   C. Organization
   D. Rumors

8. Which of the following is a typical example of behaviors that develop out of public opinion?
   A. Demands for goods and services
   B. Bureaucracies
   C. Apathy
   D. Corporate takeovers

9. Which of the following is a reform movement?
   A. Political revolution
   B. Millenarian movement
   C. Conservative movement
   D. Civil rights movement

10. The American, French, Maoist, and Russian Revolutions helped to accelerate
    A. peaceful resolution of conflicts with colonializing nations.
    B. economic and social transformation of those societies.
    C. immense growth in expressive movements existing in those nations.
    D. interest in needs of the aristocratic classes of those nations.

11. Which of the following is NOT an example of a major reform movement in recent American history?
    A. Anti-pornography movement
    B. Temperance movement
    C. Civil rights movement
    D. American labor movement

12. Which of the following best describes the successes of the environmental movement?
    A. Little success in changing public attitudes about the importance of environmental issues has occurred.
    B. Considerable progress in addressing environmental costs of economic growth occurred by the end of the 1980s.
    C. Awareness that lifestyles and habits must change in order to significantly impact environmental problems occurred in the 1970s.
    D. Clean air and water acts and beginning an environmental agency represent public concern translated into public policy.

ANSWER KEY

The following provides the answers and references for the practice test questions. Objectives are referenced using the following abbreviations: T = text, R = Telecourse Guide Reading, and V = Video.

Answers	Lesson Goals	Objectives	References
1. D	1	T1	Kornblum, p. 228
2. A	1	T2	Kornblum, p. 230
3. A	1	T3	Kornblum, p. 231
4. D	2	T4	Kornblum, pp. 232-233
5. A	2	T5	Kornblum, p. 235
6. B	2	T6	Kornblum, pp. 236-238
7. C	3	T7	Kornblum, pp. 246-247
8. A	3	T8	Kornblum, p. 247
9. D	4	V1	Video
10. B	4	V2	Video
11. A	4	V3	Video
12. D	4	V4	Video

Notes and Assignments:

Lesson 26

Social Change

LESSON ASSIGNMENT

Review the following assignment in order to schedule your time appropriately. Pay careful attention; the titles and numbers of the textbook chapter, the telecourse guide, and the video program may be different from one another.

Text:
Kornblum, *Sociology in a Changing World*,
Chapter 10, "Global Social Change," pp. 292-323.

Reading:
There is no Reading for this lesson.

Video:
"Social Change,"
from the series, *The Sociological Imagination*.

OVERVIEW

They sat on the porch of their modest home, not far from the downtown of a large midwestern city. The teenager, her pretty skin several shades lighter than the nut-brown of her grandmother's gazed at her granny lovingly. The old woman hardly had the strength to move the ancient rocker, but her voice was still strong.

"You know, girl," said the grandmother, "your great-great-grandma was a slave down in Carolina. She was a strong woman and bore seven fine children. Most of them moved off the farm and went up north to the big cities to get work."

"One of those kids was my grandpa. A hard-working man he was; worked in a factory that made tanks for the Great War. And he was learned, too. Taught himself; read every book he could get his hands on. Became a preacher in his old age and organized others of our race to fight for their rights."

"Yes, I know, Granny. Mama told me all about him, and that's why I want to be a lawyer for our people. Do you think I can make it?" the girl asked.

The aged matriarch laughed and, without missing a beat, responded, "I reckon you can do anything you want to, Felicia. Women these days are doin' all kinds of things I never would have even dreamed of. Where you goin' to school, child?"

"Harvard. Daddy says I'm so smart I can get a scholarship just about anywhere," Felicia said proudly. "But I don't have enough time to do my homework, now that I need to hold a job to bring in money since Daddy was laid off from the plant."

In these interactions you can hear hints of the major social changes that have affected societies over the past two centuries. Although it is not accurate to refer to events in individual biographies as social change, what has happened in this family is a microcosm of the broad, macro-level transformations sociologists refer to as social change.

Over the generations, this family has lived in three different parts of the country—not because they were simply wanderers, but because of economic shifts. The industrial revolution forced them to move from the farm to the city as the United States and societies all over the world became urbanized. War brought new employment opportunities, making the military a new and powerful economic force.

The fact that Felicia aspires to a profession points to changes in race relations and in the opportunity structure for minorities, including women. But now Felicia must also hold down a job, because her father is experiencing the results of another major social change: the shift from an economy rich in blue-collar manufacturing jobs to one in which smokestack industries are diminishing and high-tech enterprises are predominating.

Social change is not just another sociological concept; it is the reason for sociology itself. The science of sociology was born in the nineteenth century as a response to the tumult caused by revolutions and upheavals in France, the European colonies, and America. One of its man goals was to understand and deal with the social problems that attended the industrial revolution.

LESSON GOALS

Upon completing this lesson, you should be able to:

1. Analyze social change in its various forms and levels.
2. Analyze the modernization process and how it has altered social, physical, and cultural systems throughout the world.
3. Analyze the impact of social change on individuals and the environment, and analyze how the lives of racial and ethnic minorities have changed and have been affected by social change.
4. Analyze sociological theories of social change.
5. Analyze how social change affects individuals and how individuals help bring about change.

TEXTBOOK OBJECTIVES

The following textbook objectives are designed to help you get the most from the text. Review them, then read the assignment. You may want to write notes to reinforce what you have learned.

Text: Kornblum, *Sociology in a Changing World*, Chapter 10.

1. Define social change. Differentiate between endogenous and exogenous forces of social change, and be able to recognize examples.
2. Describe different levels of social change as analyzed by sociologists. Be able to recognize examples of these levels of social change.
3. Describe "modernization," and explain why the term should be used cautiously.
4. Explain Neil Smelser's ideas about modernization. Describe the ways that technical, economic, and ecological changes impact the whole social and cultural fabric of society.

5. Define a "developing nation." Describe the social differences between modernized and non-modernized societies. Discuss how modernization affects the world's resources.

6. Describe Andre Gunder Frank's and Immanuel Wallerstein's ideas about dependency and modernization. Identify, describe, and give examples of the three divisions of world systems Wallerstein theorized.

7. Describe how social changes have affected gender roles and the feelings of women and men. Explain why day care has become such an important consideration in this issue.

8. Identify what gains have been made and are still to be made by black Americans. Describe the observations of sociologist Cornell West.

9. Describe and exemplify the divided civic culture and its results.

10. Describe the components of the nineteenth-century evolutionary model. Discuss its two assumptions that have been criticized. Differentiate between the unilinear and multilinear models of social change.

11. Describe Pitirim Sorokin's cyclical theory of social change. Define ideational culture.

12. Describe the conflict models of social change—including Ralf Dahrendorf's theory. Discuss Dahrendorf's opinion of when social change is likely to be violent.

13. Explain the functionalist approach to social change. Be able to recognize an example of a statement that is functionalist in orientation.

VIDEO OBJECTIVES

The following video objectives are designed to help you get the most from the video segment of this lesson. Review them, then watch the video. You may want to write notes to reinforce what you have learned.

Video: "Social Change"

1. Illustrate how social change affects human relationships. Compare examples of social change that occur from outside and from inside a society or culture, especially emphasizing the Penan people and the automobile in the comparison.

2. Identify examples of both the positive and the negative effects of social change, especially changes caused by science and technology. Discuss the negative effect of social change as it relates to the invention and use of antibiotics. Using the example of the telephone lineperson, explain how social change can affect our work and our lifestyles.

3. Determine how individuals help bring about change. Be able to recognize examples. Discuss what makes a person able to maintain his or her determination to effect change.

RELATED ACTIVITIES

These activities may be used by your instructor as written assignments or as discussion topics. They may also be included as essay questions on your tests.

1. Find someone in your family or a friend's family who remembers life "back on the farm." Ask the person to describe the way it was—the dependence on subsistence farming, use of animal power, importance of the village, and so forth. Summarize the results of your investigation.

2. Describe situations or areas in the United States that meet the criteria of a peripheral region. What are the conditions that tend to make it a peripheral region? Describe them graphically.

3. What is your community doing to deal with its waste? Are there problems in coping with increasing volumes of garbage? Are there efforts to recycle? Record the results of your inquiry.

4. Are there areas of your country in which you see manifestations of all three evolutionary cycles mentioned by Pitirim Sorokin? Which cycle seems to be predominating in your country at the present time? Explain.

5. Outline a social problem that has been caused, at least partially, by changes in science and technology.

6. Discuss three technological changes that have impacted your life. Describe your initial reaction to each change. Explain the differences each technological change has made in your life—both positive and negative. (Example: Initially, I was frightened of computers. Now, I can produce papers with images, tables, title sheets, tables of contents, and spelling errors corrected by the computer. Unfortunately, I find myself spending more time on my computer and less time with my family.)

PRACTICE TEST

The following items will help you evaluate your understanding of this lesson. Use the answer key at the end of the lesson to check your answers or to locate material related to each question.

Multiple-Choice

Select the one choice that best answers the question.

1. Pressure for social change that builds within a society is referred to as
   A. hypertrophic force.
   B. entropic force.
   C. endogenous force.
   D. exogenous force.

2. Social change by definition involves all the following EXCEPT variations in
   A. ecological ordering of populations and communities.
   B. functioning of institutions.
   C. cultures of societies.
   D. political leadership through elections.

3. In a tribal society such as the Ebrie, urbanization results in
   A. new markets.
   B. a sense of anomie.
   C. villages surrounded by new neighborhoods.
   D. a confusion over expectations.

4. The term modernization should be used cautiously because
   A. implications that life in modern societies is better or more satisfactory than life in developing nations may be inaccurate.
   B. sociology has only recently begun to use the term and it is yet to be defined properly.
   C. modern societies are in a better position to care for their environment than are developing nations.
   D. colleges and universities have inadequate procedures and structures to make sure the term is properly applied.

5. According to Neil Smelser, modernization in the developing society is characterized by movement from
   A. sacred to secular.
   B. organization to disorganization.
   C. farm and village to urban concentrations.
   D. nuclear families to greater dependence on extended families.

6. Which of the following is NOT a characteristic of the modernization process in developing nations?
   A. It results in higher productivity.
   B. It tends to occur at the same rate in all nations.
   C. It increases average life expectancy.
   D. It leads to political participation by more people.

7. Although Saudi Arabia and Mexico produce large quantities of oil for the world market, both are dependent on developed nations for capital and technology.

   According to Immanuel Wallerstein's world system theory, they are
   A. developing nations.
   B. core nations.
   C. semiperipheral areas.
   D. peripheral areas.

8. Since the end of the nineteenth century, the number of married women in the labor force has
   A. decreased dramatically.
   B. remained the same.
   C. increased slightly.
   D. increased dramatically.

9. Since the civil rights movement of the 1960s, blacks in America have achieved
   A. few gains in their civil rights.
   B. equal access to jobs.
   C. gains in all our society's major institutions.
   D. equality in education.

10. One result of divided society is
    A. greater control of the political process by the government.
    B. a political system that encourages the emergence of minority parties.
    C. increased levels of educational attainment on the part of citizens.
    D. continued conflict over what, if anything, to do about the environment.

11. Which of the following is NOT an element of the evolutionary model of social change?
    A. Social change is natural and constant.
    B. Social change has a direction.
    C. Social change results from intergroup conflict.
    D. Social change is continuous.

12. Which of the following terms does Pitirim Sorokin use to refer to a cultural system that stresses sensory experiences, self-expression, and the gratification of individual desires?
    A. Materialistic culture
    B. Ideational culture
    C. Individualistic culture
    D. Sensate culture

13. Which of the following suggests that opposition among groups with different amounts of power produces social change?
    A. Cyclical model
    B. Interactionist model
    C. Functionalist model
    D. Conflict model

14. From a functionalist perspective, social change occurs as a result of such factors as
    A. population growth and changes in technology.
    B. improvements in the means of production.
    C. fluctuations between sensate and ideational periods in a culture's history.
    D. changes in the basic structure of society.

15. Which of the following is the best illustration of how social changes affect human relationships?
    A. Changes in ways young people relate to older adults
    B. Changes in technology causing people to buy new products
    C. Changes in fads that affect peer group relationships
    D. Changes occurring in the membership of music groups as music changes

16. Which of the following best illustrates the effect of social changes on American lifestyles?
    A. Telephone line persons having to work hard
    B. Telephone line persons having to learn more and be more dedicated to the job
    C. Telephone companies having to decrease rates due to competition
    D. Telephone companies having fiber optics for telephone cables

17. According to an African-American man interviewed in the video program on social change, individuals are more likely to be successful in effecting social change if they
    A. believe in the capitalistic system.
    B. retreat from society for solitude while shaping goals.
    C. work in high-density population areas.
    D. work to express the collective will and have support from faith and family.

ANSWER KEY

The following provides the answers and references for the practice test questions. Objectives are referenced using the following abbreviations: T = text, R = Telecourse Guide Reading, and V = Video.

Answers	Lesson Goals	Objectives	References
1. C	1	T1	Kornblum, p. 294
2. D	1	T1	Kornblum, p. 294
3. C	1	T2	Kornblum, p. 295, Table 10.1
4. A	2	T3	Kornblum, p. 301
5. C	2	T4	Kornblum, p. 302
6. B	2	T5	Kornblum, p. 304
7. C	2	T6	Kornblum, p. 306
8. D	3	T7	Kornblum, p. 306
9. C	3	T8	Kornblum, p. 309
10. D	3	T9	Kornblum, p. 311
11. C	4	T10	Kornblum, pp. 311-312
12. D	4	T11	Kornblum, p. 312
13. D	4	T12	Kornblum, p. 313
14. A	4	T13	Kornblum, p. 313
15. A	5	V1	Video
16. B	5	V2	Video
17. D	5	V3	Video

Contributors

We gratefully acknowledge the valuable contributions to this course from the following individuals. The titles were accurate when the video programs were recorded, but may have changed since the original taping.

SOCIOLOGISTS

Howard S. Becker, Northwestern University
Philip W. Blumstein, University of Washington
Vernon W. Boggs, City University of New York
Craig J. Calhoun, University of North Carolina—Chapel Hill
John A. Clausen, Institute of Human Development, University of California—Berkley
James S. Coleman, University of Chicago
Randall Collins, University of California—Riverside
Glenn Currier, El Centro College
Reynolds Farley, University of Michigan
Claude Fischer, University of California—Berkeley
Cheryl Townsend Gilkes, Colby College
Todd A. Gitlin, University of California—Berkeley
Felipe Gonzales, University of New Mexico
Joseph R. Gusfield, University of California—San Diego
Mildred Reed Hall, Edward T. Hall Associates
Gerald Handel, CUNY—City College and Graduate Center
Sharon Hicks-Bartlett, University of Chicago
Arlie Russell Hochschild, University of California—Berkeley
Christopher Hurn, University of Massachusetts
Paul Edward Joubert, University of Southern Louisiana
Joleen Kirschenman, University of Chicago
Karin Knorr-Cetina, University of Bielefeld
William Kornblum, CUNY Graduate School
Jack Levin, Northeastern University
Larry Lyon, Baylor University

Jane Penney, Eastfield College
Caroline H. Persell, New York University
Claire Renzetti, St. Joseph's University
Peter I. Rose, Smith College
Ruth C. Schaffer, Texas A&M University
Pepper Schwartz, University of Washington
Arthur B. Shostak, Drexel University
Neil J. Smelser, University of California—Berkeley
Robert N. Stern, ILR — Cornell University
Immanuel Wallerstein, State University of New York at Binghamton
Frank J. Weed, University of Texas—Arlington
Terry Williams, Visiting Professor—Yale University
Erik Olin Wright, University of Wisconsin
Mayer N. Zald, University of Michigan
Harriet Zukerman, Columbia University

OTHER SOCIAL SCIENTISTS AND EXPERTS

Virginia Currey, Associate Professor of Political Science, Southern Methodist University
Dr. Stan Huff, Marriage and Family Therapist, Dallas
Frances Fox Piven, Professor of Political Science, CUNY — Graduate Center
Joel Spring, Professor of Education, SUNY, College of Old Westbury

OTHER PROFESSIONALS

Roy Bode, Editor, *Dallas Times Herald*
Ron Cowart, Coordinator, Crime Intervention Program, City of Dallas
Stephen R. Donaldson, Author
Rene Durazzo, Director, Public Policy, San Francisco AIDS Foundation
Phil Dyer, President, Criminal Justice Consulting Systems, Inc.
Steve Eberhart, Music Director, KVIL-Radio
Brian Hocker, KXAS-TV, Director of Administration and Programming

Dr. Yusuf Ziya Kavakci, Director of Islamic Center, Islamic Association of North Texas, Inc.
Debra Linker, Criminal Justice Consultant
John H. McElroy, Dean of Engineering, University of Texas—Arlington
Morton Mintz, Author and retired *Washington Post* reporter
Pettis Norman, Dallas Businessman
Donald E. Paschal, Jr., City Manager, McKinney, Texas
Lewis Pollack, Vice President of Sales & Marketing, PMI McLaughlin
Jim Schermbeck, Texas United
Dr. Herbert Shore, Executive Vice President, North American Association of Jewish Homes/Housing for the Aging
Marcos Nelson Suarez, General Manager, *El Hispano News*
Linda L. Vann, Assistant Archivist, Cherokee National Historical Society, Inc.
Billy C. Wood, District Director, Office of Thrift Supervision, Dallas
Rabbi Sheldon Zimmerman, Temple Emanu-el, Dallas